Praise for *Bare...*

"I love Jeff Poppen's baref... and elegant intellectual pr... ing writing reflects a deep, personal experience in farming and gardening, as well as an informative and articulate, yet simple, interpretation of Rudolf Steiner's agriculture lectures. A must-read for adventurous agriculturalists and gardeners."

—**Luke Frey**, Frey Vineyards

"As I read Jeff Poppen's *Barefoot Biodynamics*, I could picture his grin as he talked about the changes in language that have occurred in the past hundred years since Rudolf Steiner gave his lectures on agriculture. Jeff has adapted biodynamic management to fit the specifics of growing healthy food in Tennessee. By paying attention to soil conditions from the start and documenting the responses to the same 'disturbances' over time, Jeff could respond ever-more rapidly with appropriate management using processes suggested by Steiner."

—**Elaine Ingham**, founder, Soil Foodweb, Inc.; founder and head of science and research, Soil Food Web School

"Jeff Poppen has an incredible grasp of how the cycles of nature can work together to produce the abundance that nature means for us to enjoy."

—**Will Harris**, White Oak Pastures

"Jeff Poppen is a master of self-sufficiency and community building. Without a doubt, he grows the most delicious food I have ever tasted. In this new book, he takes away a bit of the mystery behind biodynamic agriculture, so you too can grow incredibly delicious and nutritious food."

—**Sean Brock**, James Beard Award–winning chef and author

"*Barefoot Biodynamics* highlights the contrast between natural ecosystem processes and today's prominent model of trying to 'manipulate' nature. An excellent read for all who want to understand the keys to growing nutrient-dense food."

—**Gabe Brown**, author of *Dirt to Soil*;
co-founder, Understanding Ag

"Jeff Poppen's *Barefoot Biodynamics* is a heartfelt and matter-of-fact account of the journey of a human being, a farm, and a community during times of dramatic change in agriculture. Especially impressive is the final chapter, in which Rudolf Steiner's complex *Agriculture Course* lectures and discussions are put into plain language. Simultaneously brilliant and humble: Well done, Jeff Poppen!"

—**Mark Shepard**, author of *Restoration Agriculture*
and *Water for Any Farm*

"In *Barefoot Biodynamics*, Jeff Poppen illuminates the message of Rudolf Steiner's *Agriculture Course*, not just by distilling Steiner's words, but through himself being a result of the course. Jeff is a farmer humbly living the lectures, in the wonder of them, the awe. Jeff pays homage to the farm organism, the tillage, the cover crops, the festivals, the cattle, the food, the bounty, and yes, the heartaches, the failures, the struggles."

—**John Peterson**, founder, Angelic Organics; subject of
the documentary *The Real Dirt on Farmer John*

"Brilliant and fantastic! *Barefoot Biodynamics* is delightful, fun reading that fills us with joy, while also imparting deep insight and wisdom into gardening, farming, and the natural world."

—**Kristina Villa**, co-executive director,
The Farmers Land Trust

"With *Barefoot Biodynamics*, Jeff Poppen brings biodynamic agriculture to the service of all. With enthusiasm and great humor, Jeff translates Rudolf Steiner's esoteric recommendations into useful agricultural approaches."

—**Bryan O'Hara**, author of *No-Till Intensive Vegetable Culture*

Barefoot
Biodynamics

Also by Jeff Poppen

Agriculture Abridged: Rudolf Steiner's 1924 Course
The Barefoot Farmer: The Best of the Barefoot Farmer
The Best of the Barefoot Farmer, Volume II

Barefoot Biodynamics

How Cows, Compost, and Community
Help Us Understand
Rudolf Steiner's *Agriculture Course*

Jeff Poppen

Foreword by
Sandor Ellix Katz

Chelsea Green Publishing
White River Junction, Vermont
London, UK

Some of the material in this book is from *Agriculture Course* by Rudolf Steiner. Published by Rudolf Steiner Press 2004. Translation © Rudolf Steiner Press, 1958. Reproduced with permission of Rudolf Steiner Press through PLSclear.

Some of the material in this book is adapted from *Agriculture Abridged* (Jeff Poppen, 2021).

Project Manager: Natalie Wallace
Editor: Fern Marshall Bradley
Copy Editor: Will Solomon
Proofreader: Laura Jorstad
Indexer: Shana Milkie
Designer: Melissa Jacobson
Page Layout: Abrah Griggs

Printed in Canada.
First printing April 2024.
10 9 8 7 6 5 4 3 2 1   24 25 26 27 28

Our Commitment to Green Publishing

Chelsea Green sees publishing as a tool for cultural change and ecological stewardship. We strive to align our book manufacturing practices with our editorial mission and to reduce the impact of our business enterprise in the environment. We print our books using vegetable-based inks whenever possible. This book may cost slightly more because it was printed on paper that contains recycled fiber, and we hope you'll agree that it's worth it. *Barefoot Biodynamics* was printed on paper supplied by Marquis that is made of recycled materials and other controlled sources.

Library of Congress Cataloging-in-Publication Data
Names: Poppen, Jeff, author. | Katz, Sandor Ellix, 1962- author of foreword.
Title: Barefoot biodynamics : how cows, compost, and community help us understand Rudolf
   Steiner's Agriculture Course / Jeff Poppen ; foreword by Sandor Ellix Katz
Description: White River Junction, Vermont : Chelsea Green Publishing, [2024] |
   Includes bibliographical references and index.
Identifiers: LCCN 2023059597 (print) | LCCN 2023059598 (ebook)
   | ISBN 9781645022480 (paperback) | ISBN 9781645022497 (ebook)
Subjects: LCSH: Steiner, Rudolf, 1861-1925. Geisteswissenschaftliche Grundlagen
   zum Gedeihen der Landwirtschaft. | Organic gardening. | Biodynamic agriculture. |
   Handbooks and manuals.
Classification: LCC SB453.5 .P66 2024 (print) | LCC SB453.5 (ebook) |
   DDC 635/.0484—dc23/eng/20240129
LC record available at https://lccn.loc.gov/2023059597
LC ebook record available at https://lccn.loc.gov/2023059598

Chelsea Green Publishing
White River Junction, Vermont USA
London, UK

www.chelseagreen.com

MIX
Paper from
responsible sources
FSC® C103567
FSC
www.fsc.org

In appreciation of Dr. Rudolf Steiner

Contents

Foreword

Jeff Poppen has been farming for fifty years, since he was a teenager. By the time I arrived in Tennessee in 1993, he was already iconic there, widely known as "the barefoot farmer," freely sharing vegetables at public events, and evangelizing to anyone who would listen about local, organic, and biodynamic agriculture. I've always found him generous with his insights and wisdom, and over time we have become friends.

Whenever I visit Jeff's farm in Red Boiling Springs, I experience a powerful sense of abundance. The vegetables are so big and look so healthy and vibrant, and there are so many of them. Every year, Jeff allows me to harvest a truckload of fall radishes and cabbages to ferment, and it barely makes a dent. Jeff talks about the vegetables as a by-product of his primary crop: good soil. That's one of the important ways in which he and I have found so much common ground. Building soil fertility involves cultivating microorganisms, just as fermentation does.

In this book, Jeff lays out the methods he uses and the ideas that underlie them. Like most farmers everywhere, Jeff is eminently practical, and learns primarily through observation and experimentation. He's also read through many old farming books, which inform his thinking. Jeff has been especially influenced by the early-twentieth-century Austrian philosopher Rudolf Steiner, whose lectures on agriculture inspired the biodynamic farming movement.

Jeff readily admits that some of Steiner's ideas may seem strange. At the beginning of his biodynamic journey, Jeff was told: "You don't have to believe it; just try it." Jeff has kept at it for decades now, based on results. He makes his own soil-building preparations as prescribed by Steiner (and described in this book), and every year when I visit him, he gives me some of the horn manure preparation to use in my garden.

The horn manure looks like a small scoop of compost. The thrilling magic of it is revealed only in the somewhat laborious process of activating it, by mixing it vigorously for an hour into water, forming a vortex that sucks air into the solution and enables the aerobic organisms to proliferate. I add the horn manure to a 5-gallon bucket about half full with lukewarm water. Then, using a stick, I stir rapidly in circles around the edge to create the vortex, with the edge rising and the center descending, that pulls in the air. I stir for a minute or two in one direction; then reverse; and so on for an hour. This is a task that would be impossible for me alone, but has been great fun to do with friends. After the first ten or fifteen minutes, the solution is fizzing and hissing, demonstrating in real time the effectiveness of our efforts. After an hour, close to sunset, we dip the bristles of an old broom into the horn manure brew and fling the liquid to disperse it onto the garden beds, spreading this teeming biodiversity to settle into the soil with the evening dew.

Beyond using the preparations Jeff gives me, I have not followed Steiner's methods. Like Jeff, I am an iconoclast, not prone to accept the singular truth of any particular book, system, or person. Yet, as someone who has been lucky enough to visit many farms around the country and around the world, I have observed that the biodynamic farms have been some of the most stunning, verdant ones.

One focus of this book is Jeff's interpretation of Steiner's *Agriculture Course*, upon which biodynamic agricultural principles and practices are based. Jeff understands and explains Steiner's ideas through the prism of his own experiences, but ultimately, this book is a much broader compendium of Jeff's thinking, based on his observations over a half century of farming. Throughout the book, Jeff emphasizes that the farmer must be a keen observer, continual student, and analytical theoretician. Ideas such as Steiner's can be very helpful "guiding lines," as Jeff puts it; however, it is imperative to reconsider and revise ideas based on observation and lived experience. As Jeff writes: "Far from following recipes, we must figure out, on our own unique farms, how to use the guiding lines Steiner offered to find our own individual way."

Jeff's reflections about farming are down-to-earth folk wisdom. He's mentored many young farmers, and with this book he's sure to inspire many more. I know he has inspired me. Learn from Jeff, and learn from Steiner and Jeff's other guiding lines. But as Jeff suggests, use the ideas to sharpen your observational skills. Our greatest teacher is experience, and wisdom is readily available from the plants, animals, soil, and microbes that we work with.

SANDOR ELLIX KATZ,
author of *The Art of Fermentation*
and other fermentation bestsellers

Introduction

R eading Rudolf Steiner reminds me of the saying, "You can't believe everything you read." A few years after starting an organic farm in Tennessee, I read my dad's copy of Steiner's *Agriculture, A Course of Eight Lectures*, in which Steiner discussed principles and practices that later became known as the biodynamic method. I immediately realized two things. The sentences and ideas I understood seemed extremely important and relevant, but others made no sense to me at all. I've always lived on a farm, so I could tell this guy knew something about farming. But so much of what he said seemed so far out, I could see why the method wasn't more popular.

I visited some biodynamic farms and was really impressed. The gardens were beautiful, the soil was rich and alive, and the farms had a radiant ambience of health. But I figured anyone who paid that much attention to all the details involved was bound to be a good farmer anyway. Told I didn't have to believe in biodynamics for it to work, I tried the method for a year, saw great results, and have been trying to figure out what it's all about for the last forty.

Dr. Rudolf Steiner, 1861–1925, wrote twenty-five books and gave around six thousand lectures on a wide range of subjects. His parents were true children from the glorious Lower Austrian forest region, and Steiner learned from the stories he heard told by peasants while he rambled about in the forests. Geometry fascinated him at an early age, and he got great satisfaction knowing that a person can live within the mind, in the shaping of forms perceived only within oneself, entirely without impressions upon the external senses. As he grew up he sought to prove that spirit is the agent in human thinking. He hoped that someday he could blend natural science with the knowledge of the spirit.

No one was interested in Steiner's inner spiritual perceptions until he met an "uneducated" herb gatherer near Vienna. His time with this initiate led to a theory of knowledge without the limits set by Kant and other philosophers. It eventually became his doctoral thesis, *The Philosophy of Freedom*, published in 1893.

During his lifetime, Steiner studied everything under the sun, and maybe beyond it, too. He attended the Technical Institute of Vienna, where he studied math, chemistry, and biology. Besides herbal and homeopathic medicine, folklore, and Indigenous cultures, his curiosity led him into Eastern religions, ancient mysteries, and many branches of natural science. He developed his natural clairvoyance to a remarkable degree, eventually tarnishing his highly respected reputation as a leading intellectual lecturer in European academia by admitting this publicly.

Steiner delivered his lectures in German, and they were later translated from unrevised shorthand reports. Unlike books in which the concepts have been worked out and revised, Steiner's lecture courses capture an in-the-moment experience. The audience was familiar with his other books and lectures, so they were acquainted with his unique terminology.

My dad's copy of Steiner's agriculture lectures was the Adams translation (originally published in 1958, and since re-released with the title *Agriculture Course: The Birth of the Biodynamic Method*). But even after studying other works by Steiner, I found it difficult to comprehend the ideas in these lectures. In 1987, I helped form the Southeast Biodynamic Association with my late mentors, Hugh Lovel, Harvey Lisle, and Hugh Courtney. We read from the lectures at every annual SBA conference, and often in between. In 1993 a new translation came out, written by Catherine E. Creeger and Malcolm Gardner, which helped broaden our perspective. I also began writing a weekly newspaper column in 1993 for the *Macon County Chronicle* and I called it "Small Farm Journal." After six months, my editor renamed it "Barefoot Farmer," and I've published two collections of articles under that name.

I'd learned that putting concepts into my own words helped me understand them better, so I began writing out my own interpretation of the agriculture lectures. Many others felt the need for a simplified version and encouraged me in my effort to interpret Steiner into everyday language.

My farming has been affected by what I've learned from many other books, farms, and people, but I can always open the *Agriculture Course* and find a new insight. It is profound, enigmatic, and the cheapest farming method I've found. The longer I farmed, the more I understood the common sense underlying these enigmatic lectures.

In 2012, this project became my wintertime focus, and for the next few years I worked on it during the off-season. My friendship with Hugh Lovel blossomed into many fruitful farming discussions, resulting in his offer to help edit the lecture-by-lecture overview I was writing. He loved the idea of creating a grade-school textbook from the course. I self-published our efforts in 2020 as a small book called *Agriculture Abridged*. Folks have been trying to decipher Steiner's agriculture course for a hundred years, and this booklet was simply one more effort.

We felt the need for a publishing house to distribute it more widely. Chelsea Green Publishing showed an interest, and editor Fern Bradley and I decided that adding stories from my own farming experiences might be of help. This book is the result, and some of the text from *Agriculture Abridged* is included as the final chapter, "*Agriculture*, Simplified." It was a pleasure to work with the Chelsea Green staff, and I can't thank them enough for their support, especially since I am computer illiterate.

Although there are many important concepts in the agriculture lectures, I have organized the opening chapters of this book around eight concepts that have been fundamental to my farming. I started with the insight that nitrogen is a result of agriculture and not an input. The self-sufficient farm is also a dominant theme in the lectures. I then wanted to highlight Steiner's less serious side and the need farmers have for community. Much of what is wrong in agriculture today can be traced to following the wrong advice, and I wanted to emphasize that I listen to farmers, not professors and economists. The next chapter is about the influences of the atmosphere, particularly the "personalities" of carbon, oxygen, nitrogen, hydrogen, and sulfur. Then I discuss the terrestrial elements, the minerals in the earth. Steiner frequently implores us to put our plants in a soil rich in humus, and I certainly concur. The biodynamic preparations are one way of enhancing good farming practices, and it took me a long time to learn how to make them well and grasp their effects, so I devoted a chapter to this topic, too.

Once I finished writing those eight chapters, I realized I had skipped over other important guidelines, so I introduced a few more in the chapter called "Guiding Lines."

Steiner advised farmers to go on manuring as before, and to understand what he meant by "before," I decided to study old farming textbooks. The chapter with this title covers some of the invaluable lessons I learned from those old books about how folks farmed in the late nineteenth century. A chapter on Goethean science follows, to help us learn how to learn from nature. Then we come to the heart of the book, the lecture-by-lecture overview of the agriculture course. Together, Hugh Lovel and I had a combined seventy-five years of studying the agriculture course, and our mentors, Harvey Lisle and Hugh Courtney, had a combined one hundred years of studying it, so Lovel was the perfect fit to help edit my *Agriculture Abridged* booklet. He was an open-minded scientist who had a true love for farming, and his collaboration kept the chemistry correct.

I certainly love farming, too. Most of the rows on our cropland are on 44-inch centers for ease of cultivation. This season the potato and winter squash fields amount to 3 miles of rows each, the sweet potatoes, pepper, and tomato patches amount to 2 miles in total, and the onion, corn, beans, cucumbers, and watermelons make up about a mile of rows. Besides these 9 miles, we also grow about an acre of fruit, flowers, herbs, and a kitchen garden.

The potatoes and corn haven't required any hoeing, due to appropriate harrowing, cultivating, and hilling with the Farmall 140 tractor. We hoed out the winter squash, melons, beans, and cucumbers once, and the peppers and sweet potatoes twice, before hilling them. The onions needed four hoeings, the 1,500 row feet of garlic is mulched, and all the crops are laid by and looking good.

For several years the farm grew twice this amount for a two-hundred-member CSA and other markets. I initiated and managed five different 1-acre gardens in Nashville for a few years, too. I had three hired hands then instead of one part-timer as I do now, and last year I made the tough decision to retire the CSA and sell the farm's produce wholesale. I'm happy to let a younger generation do the hustle and bustle of marketing, and volunteers still come to the farm to help with some of the bigger jobs like planting, hoeing, and harvesting.

The 5 acres or so of cropland received about 150 tons of beautiful, black biodynamic compost, along with a few tons of lime, lots of cover crops, and the biodynamic preparations made with manure and silica that were buried in horns. These gardens will yield around 60,000 pounds of vegetables, produced without using irrigation, plastic mulches, or hoop houses. We'll also put up around 125 rolls and 200 bales of hay, which will keep a few dozen head of cattle, their calves, and the dairy goats happy throughout the winter. I'll turn the resulting wastes into compost for future fields and gardens.

The biodynamic method rests on the principle of integrating the appropriate ratio of livestock and legumes to cropland so fertility is maintained without having to buy stuff. That means we try to grow all our animals' food, as well as our own, and move our animals and crops around enough to keep the fields fertile and full of humus. The question begs itself, "Would the farm be as abundant if I did everything else but left out the biodynamic preparations?" I really don't know the answer. Maybe making the preparations helps with another important task, the development of the farmer's intuition. After all, what we're growing, ultimately, is ourselves.

CHAPTER ONE

Secrets of Manuring

*B*arefoot in the title might seem misleading. The
word doesn't appear anywhere else in the book,
but it implies my simple, down-to-earth approach. This
book is about biodynamic gardening and farming and
how I've made my livelihood since the mid-1970s. Cows,
compost, and community are fundamental. A hundred
years ago, and long before, farming was taught through
experience and was more or less "organic." As chemicals began to be used, a breath of common sense came
through Rudolf Steiner's lectures on agriculture in 1924,
with his admonishment to avoid artificial fertilizers and
rely on cows, compost, and traditional methods instead.
The instinct of how to grow plants and animals lies dormant in us and can be inspired to awaken. The advice
given in the course, which Steiner referred to as guiding
lines, became the foundation for our farm, and is also the
foundation of this book. Although Steiner's esoteric ideas
are fascinating, I've tried to keep this book easy to read.
I'll leave it to you to dig deeper into Steiner, biodynamics, and your garden. You can take your shoes off or not.

After returning home from delivering the agriculture course, Steiner gave a report to his colleagues

about what happened. Lectures began at eleven in the morning and lasted until one in the afternoon, and were followed by a midday meal and roaming the grounds of the estate. About a hundred farmers and scientists attended. Afterward, they developed their farms and research from the guiding lines he gave.

> *With regard to the agriculture course, the first consideration was to outline what conditions are necessary in order for the various branches of agriculture to thrive. Agriculture includes some very interesting aspects—plant life, animal husbandry, forestry, gardening, and so on, but perhaps the most interesting of all are the secrets of manuring, which are very real and important mysteries . . . during the last few decades the agricultural products on which our life depends have degenerated extremely rapidly.* *

In Steiner's time, "manuring" meant how farmers fertilized the soil for growing crops. He was well aware of what was causing the degeneration of farm products. It was the new soluble artificial fertilizers. The secrets

---

* This quotation is from the translation by Catherine E. Creeger and Malcolm Gardner published by the Bio-Dynamic Farming and Gardening Association, Inc., in 1993. All other Steiner quotations in this book are from the George Adams translation first published by Rudolf Steiner Press in 1958.

9

of manuring he refers to are the living interactions of microorganisms within the soil. The activity of bacteria, fungi, and other microbes in soil, compost, and manure are indeed real and important, but they were mysterious back then because so few people understood microbial interaction in the soil. Steiner understood, though, and could discern the detrimental effects of synthetic NPK (nitrogen/phosphorus/potassium) fertilizers on those microbes.

The importance of diverse, balanced microbial activities is now well established. We need our microbial partners in our stomachs and on our skin. Animals need them, plants need them, and so does the soil. Their presence in the soil helps ensure healthy plant growth. An immediate halt to chemical fertilizing and returning to the use of compost instead would turn degeneration into regeneration. Steiner understood the significance of manure, and how it should be handled. He grew up surrounded by farms that depended on manure to keep the fields fertile.

* * *

The small farm where I spent my childhood was a nursery and landscaping business, and my parents were plant lovers. My dad had retired early from careers in math and psychology, bought a farm outside of Chicago, and developed a homestead with my mom. They were following a dream inspired by Helen and Scott Nearing's book *Living the Good Life*, about two professors who went back to the land. Mom and Dad would get such joy out of a new plant blooming, tending the perennial beds, or finding the first spring wildflowers. This love must have rubbed off on me.

After I moved away from my family's farm to start my own in Tennessee, I wanted to identify every plant in the fields and forest, and experience how to use each for food and medicine. Local people shared what they knew about the folklore that abounded in rural Appalachia. These folks didn't speak like they were well educated, but I quickly came to appreciate how intelligent they were. They had a love and knowledge of plants and animals deeper than any book could convey.

Debby, my partner in this adventure for the first twenty-five years, and I both grew up with animals on our childhood farms. Our parents believed farms needed both crops and livestock. Debby's family milked twenty Holstein cows by hand twice a day, besides farrowing a few hundred hogs and growing 200 acres of grain. If grain prices were high, they sold grain and feeder pigs. But if grain prices were low, they could use the grain to feed the young pigs and sell hogs later. My family raised hundreds of chickens every year. We sold eggs, and if someone came to the farm to buy a chicken for dinner, Mom would put water on to boil, pick up a bird and wring its neck, and pluck and clean it then and there. When Debby and I first moved to Tennessee, it seemed natural to get a milk cow, horses, chickens, and a pig. Animal husbandry became integral to our farm. I won't say I love livestock all the time, but I can't imagine farming without animals.

Surrounded by the beautiful Tennessee forests, I became even more of a tree lover. The woodland flora and fauna fascinated me. I let some cleared land grow up into woods and began making forest trails to special places. Our small nursery turned into the grafting of old varieties of fruit trees. In the recent past every farm had an orchard, and I loved meeting people who liked to share stories about the old days. Not many still knew how to graft, so I got to save their old family heirlooms while learning a bit of local history.

Soon gardening became my passion and I wanted to grow everything possible. Through groups of seed savers, and fruit and nut explorers, along with old-time local gardeners, I exhausted every variety I could find. I grow the varieties I grow now because of what I learned during those experimental years. We mulched, double dug, and practiced permaculture, and I grafted, propagated, and pruned anything I could get my hands on. Just learning how to work a full day was an accomplishment, as I still had those lazy bones common among teens. We were learning how to grow and preserve our own food, make hay, and tend cattle. I began selling vegetables from pretty gardens that expanded annually, and good-quality compost was making a noticeable difference.

When I first read the passage above about secrets of manuring, I wondered what they were. How in the world could manuring be more interesting than plants, animals, forestry, or gardening? I enjoyed making compost and seeing what it could do for our depleted soil, but it didn't seem that interesting or all that secret. Why was food quality degenerating extremely rapidly by the 1920s? The answers to these questions came slowly, gaining momentum the more I learned about microbiology and biochemistry.

Steiner's early training in mathematics, chemistry, and biology gave him a solid foundation for understanding how chemistry works in life processes. Of particular interest is nitrogen, how it moves in and out of the soil, and the microbes that can transform atmospheric nitrogen into forms that are available to plants. Artificial fertilizer began to be used in the late nineteenth century based on the scientific research of Justus von Liebig, the father of modern chemistry.

Liebig discovered which elements were necessary for plant growth. He originally believed and taught that they needed to be restored to the soil in a water-soluble form, although he later changed his mind. Farmers were learning that although applying soluble fertilizer to soils made plants grow lushly, the soil eventually became less fertile and problems grew more prevalent. But if they also used manure, compost, and cover crops, the soil and plants stayed healthy. Scientists discovered what farmers already knew—the importance of humus—by observing the little wiggly things in their microscopes.

* * *

Debby and I were part of a group of organic farmers numbering fewer than four thousand in the United States in the 1970s. This small group helped research organic methods and develop organic markets, and began forming statewide associations to host conferences. Although organic farming was ridiculed as a fringe idea that could never feed the world, we realized the dangers of agricultural chemicals and knew there had to be a better way. Organics worked for us, and the methods we used had fed the world before the introduction of chemicals in farming.

I gradually learned a bit of history, which helped me understand why these chemicals were being promoted so widely, and why organic farming was getting such bad press. Controlling the manufacture of ammonium nitrate meant controlling the production of both food and weapons, and that was a powerful place to be.

Seventy-eight percent of the atmosphere is nitrogen, but plants can't access it directly. It's an inert gas that needs to be transformed so plants can absorb it through their roots (plant roots can't absorb gaseous nitrogen). Lightning can transform atmospheric nitrogen, and so can soil life.

It helps to keep in mind that plants take in nutrients such as nitrogen in two ways. One way requires the assistance of soil-dwelling microbes; the other way is to put the nutrients into the soil in a soluble form so when the plant takes in water, it also takes in the nutrients. The former was developed by nature, the latter by humans.

Before 1914, nitrate for making gunpowder and fertilizer was mined in caves as saltpeter (potassium nitrate), the final product of decaying bat guano, and also from large reserves of sodium nitrate from bird guano off the coast of Chile. A decisive naval battle was fought there at the beginning of World War I, after which the loser would be hard-pressed to mine enough saltpeter to supply their armies with enough gunpow-der. England beat Germany, but German scientists discovered how to synthesize ammonium nitrate from the atmosphere. Weapons factories sprouted up throughout Germany to supply their army, prolonging the war until 1918. After the war these factories began to produce fertilizer, changing agriculture dramatically.

* * *

No longer did farmers have to raise livestock and spread manure or grow large crops of legumes just to plow back in. Agricultural education also took a 180-degree turn, separating crop production from animal husbandry, which was impossible before then. The use of artificial nitrogen fertilizers increased grain yields worldwide dramatically. Within a hundred years, these grain crops were directly responsible for an extra two and a half billion people, or about a third of the Earth's population. The widespread use of artificial fertilizer has also been accompanied by social, economic, and ecological consequences we feel acutely today.

But using nitrate fertilizer somehow made the soil unable to grow crops if you quit using it, so selling fertilizer became an extremely profitable, repeat business. Scientists who understood biochemistry, such as Steiner, began to figure out that soluble nitrogen in the soil was destroying the nitrogen-fixing microbes, which thrive in soil supplied with manure, compost, and cover crops. When humus was restored, the valuable microbes returned. Naturally grown plants proved to be healthier, better flavored, and did not impoverish the soil. It was becoming clear why food quality had degenerated. When food products were grown with nitrogen salts for fertilizer, their nutritional qualities degenerated extremely rapidly. The adage "Nature bats last" means that she'll be the final determining factor, so it's no surprise to find that her way of maintaining soil fertility, secret or not, is far wiser than what humans came up with.

Steiner insisted that the nitrogen for our farms should come from our livestock and legume crops. He believed that a farm is healthy only inasmuch as it provides its own fertilizer from its own stock, and that the only sound fertilizer is cattle manure. The word "cattle" can mean any of the domestic ruminants, such as goats and sheep, too. He called legumes "the lungs of the earth" and divided the entire plant world into two when contemplating it in relation to nitrogen. The use of legumes is essential on the farm to have this breathing in of nitrogen, which is accomplished by legumes' symbiotic relationship with certain soil bacteria. Other plants have other roles.

Plants release many of the sugars they make (via photosynthesis) into the soil as exudates from their roots. A diverse range of bacteria and fungi live near the roots of plants and they ingest the sugary exudates to fuel their reproduction. Protozoa in the soil eat bacteria and fungi and excrete amino acids that contain nitrogen. Plants can take up those amino acids and make use of the nitrogen therein. Many other living beings are involved in these complicated processes as well, to the tune of billions in every spoonful of soil.

When the leaves of plants wilt, they take in water, preferably free from adulteration, through their roots.

If the soil water contains ammonium nitrate from chemical fertilizer, plants have no choice but to take up that nitrate as they drink. But this nitrate must then be converted into nitrogen in amino acid form, and that requires energy. The plant must use some of its stored sugars to fuel the conversion. This is why chemically fertilized plants don't taste as sweet, and also why they become susceptible to pests. Insects have trouble digesting the complex sugars and amino acids in healthy plants. Thus, insects will not feed on plants that have a high sugar content.

The agriculture course is a direct response to the discovery of the process for synthesizing artificial nitrogen that in turn led to the new fertilizer industry. Speaking with a chemist from the facility where nitrogen synthesis was first developed, a few years before giving the course, Steiner mentioned that mineral fertilizers could be used, but the types of salts added to the soil had to be better chosen than was usually the case. Above all, Steiner emphasized, no nitrogen salt should be used. He recommended kainite for use as a fertilizer. Kainite is a naturally occurring rock dust that contains sulfur, magnesium, and potassium—elements that can help feed the valuable nutrient-fixing microbes in a humus-rich soil. A mantra in the organic movement is to feed the soil, not the plants.

Steiner looked at the bigger picture of what is best for human nutrition, explaining that plants that get nitrogen and other nutrients as a result of microbial interactions with the air and minerals in a farm's soil were clearly better for human health. Very few people had access to this level of biochemical research, though, which is one reason Steiner referred to this knowledge as "secrets which were very real and important mysteries." But the real secret is the self-contained farm organism where the process of high-quality composting recycles the farm's own atmospheric elements. The protein in our bodies that enables our thinking and consciousness comes from the nitrogen, carbon, oxygen, hydrogen, and sulfur in the farm's own air. It is brought to life by the farm's crops, particularly legumes and domestic animals, along with valuable additions from soil minerals and microbes, forests and birds, shrubs and mammals, wetlands and mushrooms, and the far reaches of our imaginations. This is the essence of good farming, and I knew I needed a self-sufficient farm.

— CHAPTER TWO —

A Self-Sufficient Farm

A farm needs cattle, Dad informed us. It was 1974, and Debby and I had just settled into our Tennessee homestead, which had evidently revolved around livestock in the past. Dad had witnessed the deterioration of the soil and local rural economies in the Midwest during the previous decades and attributed it to the removal of livestock from cropland. So we got a herd and spent the next several decades chasing them back into the pastures. They were trying to teach us rotational grazing.

The young farmers received more fatherly wisdom. Don't buy anything, just grow what you need on the farm. This was easy to follow given our income level. I was enamored with Steiner's *Agriculture Course* when I recognized similar advice there.

> *Whatever you need for agricultural production, you should try to possess it within the farm itself (including in the "farm," needless to say, the due amount*

of cattle). . . . A thoroughly healthy farm
should be able to produce within itself all
that it needs.

The ideal of living on a small, self-sufficient farm became a goal of mine during my late teenage years. Affected by the cultural revolution of the 1960s, many felt big changes were inevitable. The political and economic system wasn't fair, was hurting the environment, and was further dividing the haves from the have-nots. The rich were getting richer and the poor were getting poorer. I didn't want a job, because paying income taxes meant I would be supporting the war in Vietnam. It was clear that the counterculture would need a land base to realize its goals of a more just and peaceful society. The need to be independent and take care of ourselves led us to buying an abandoned farm way off the beaten path in very rural Tennessee.

There was no electricity, the water source was a spring, and the cabin was a simple shack built in 1910. We cooked on an old Black Diamond wood-burning cookstove and heated the cabin with a wood heater. Lily supplied our milk, a pig our meat, and chickens our eggs. We bought wheat for two dollars a bushel from a neighbor and ground it into flour in a Corona mill to make bread and dog food. All we needed now was vegetables and fruit from a big garden like our parents had grown every year.

This proved to be more difficult on our new land than on the rich, black soil south of Lake Michigan. A shovel would not penetrate the heavy clay. It took a pick to break apart the soil on the hill by the cabin. Down in the bottomland, near the creek, the soil was better and grew a great garden. You can imagine our chagrin one rainy morning when we awoke to see the whole bottom underwater. I'd heard of flash floods, but had never seen one, and it was impressive. Back to gardening on the hill. We began cleaning manure out of our neighbors' barns to make compost with, supplementing what we gathered from our own small herd.

We were an anomaly, kids wanting to follow a lifestyle common to the elders in our community, elders whose own children had gone north to pursue work and the lifestyle we had left behind. We broke a horse to work, tended our big garden, and canned beans and tomatoes. We rebuilt an old root cellar, which we filled with potatoes and homemade wine. Generally speaking, if something required money, we could do without it.

This relinquishing of unnecessary things seemed to be a prerequisite for a self-sufficient farm. The farm became our source of food, fuel, and entertainment. Other "back-to-the-land" folks were moving onto farms nearby, which encouraged me to start a small business grafting fruit trees and raising dairy cattle. We felt part of a movement, and the tides to come would need orchards and milk cows. But many of these new folks were also new to farming and did not last long. By 1980, nobody seemed to remember there was an anti-materialism revolution going on. I switched to selling vegetables and beef cattle.

To get away from using kerosene lighting, we traded for some used solar panels in 1981. A junked Dodge Dart supplied a battery, regulator, and taillights, and we had 12-volt lighting in our cabin. For hot water I wrapped a copper pipe around the cookstove pipe and a piece of metal around that to make a heat exchanger. By placing a water tank above and behind the stove, cold water naturally sank to the level of the heat exchanger, got warm, and then rose to the top, where we could draw off the hot water.

After a bit of fiddling around, I could usually get Dad's old 8N Ford running long enough to plow the garden or bush hog a meadow. Our cattle herd grew to about two dozen, and I made and spread a lot of compost with a pitchfork. Neighbors helped put up hay, and I began to understand the efficiency of *local dependency* over self-sufficiency. When I encountered Steiner again a few years later, I smiled at his suggestion that a farm is true to its essential nature, in the best sense of the word, if it is conceived as a kind of individual entity in itself—a self-contained individuality. Now that I had learned how difficult the self-contained approach could be, I really appreciated his next statement, that this ideal cannot be absolutely attained, but should be observed as far as possible. Well, at least he gave us some wiggle room.

Years later I learned that the farm where Steiner's lectures took place included 18,000 acres and hundreds of people doing all the different jobs that make farming and a community possible. It takes millers, blacksmiths, butchers, bakers, and candlestick makers, among many others, to help create the ideal self-contained entity we call a "farm." Our farm was a long way from that ideal.

Whether the introduction of Dad's big Ford 600 tractor, in 1987, and the small garden tractor a few years later led in the direction of a self-sufficient farm is an open question. With more horsepower available, I sold more farm products, but then there were the costs of running the equipment and driving trucks to haul the bigger harvest to markets. I liked having the ability to manage the farm better, but still spent a lot of time splitting wood for both stoves, putting up our food, and tending to all the other homesteading chores.

Walking the perimeter fence lines, I felt responsible for all that lay inside them. My mind was continually engaged with what went on within our borders. I thought constantly about crops, erosion, roads, fences, animals, and compost, and was a bit of a bore when talking to non-farmers.

By the turn of the century, I had switched to a gas range for cooking, creating less self-sufficiency but freeing up more time for developing the farm. The gardens grew a bit in acreage each year, and eventually I had several old tractors, the appropriate antique machinery, and 10 acres in vegetable production. The farm certainly developed a unique individuality, an expression of my own. Although less self-contained, it was having a rippling effect outside our borders as people became inspired by a biodynamic community farm.

Overwhelmed with so many irons in the fire, I took special interest in two books by Louis Bromfield.

In *The Farm*, he described life on his grandparents' self-sufficient farm in Ohio near the end of the nineteenth century. Covering everything from dairy, meat animals, and workhorses to gardens, orchards, and grain fields, *The Farm* is a beautiful account of the farm life that Steiner's generation grew up around. Bromfield's attempt to replicate this model fifty years later, detailed in his book *Malabar Farm*, failed. He found it unreasonable to try to do everything his grandparents had done and narrowed his focus to just a few farm enterprises. Ultimately, this was also our experience at Long Hungry Creek Farm.

We still grow and preserve most of our own food and grow the feed for our animals. Our farm has an individuality, a uniqueness, and a diverse set of crops and landscapes. But I need neighbors for companionship, community, and help with the jobs I'm not particularly good at. I have friends who keep bees here, help fix machines, and do carpentry projects. As Steiner said, "Make the farm, as far as possible, so self-contained that it is able to sustain itself. As far as possible—not quite! Why not? In outer life, within our present economic order, it cannot be fully attained. Nevertheless, you should try to attain it as far as possible." In other words, don't let perfect get in the way of good. Then I read Steiner's basic social axiom, which both confused and comforted me.

> *The well-being of an entire group of individuals who work together is the greater, the less individuals claim the income*

> *resulting from their own accomplishments for themselves, that is, the more they contribute this income to their fellow workers and the more their own needs are met not through their own efforts but through the efforts of others.* *

The more a farm could be self-contained as far as producing its own fertilizer, the better it would be, but interaction with the larger community was necessary for society as a whole. If I eat my own corn and potatoes, I can be a hermit. But if I eat your corn and you eat my potatoes, the resulting interaction fosters our social well-being.

Community Supported Agriculture (CSA) programs sprouted directly out of Steiner's socioeconomic ideas, and the CSA model offered the farm a steady outlet for our production. I immediately jumped on the CSA bandwagon in 1988. A key realization was that I loved working with nature—the plowing, composting, planting, and tending of crops and animals. It feels honorable to be producing quality food with few off-farm inputs. But distribution and marketing were better left to folks who liked working with the public. So I hired out deliveries and the CSA management and stayed

---

* This version of the axiom is as it appears in *The Fundamental Social Law: Rudolf Steiner on the Work of the Individual and the Spirit of Community* by Peter Selg, translated by Catherine E. Creeger (SteinerBooks, 2011).

on the farm. The customers supported everything the farm did, so my farming decisions were based on what was best for the long-term health of the farm, not just the economic side.

I never went to farmers markets, because I wanted to be free to go to parties and gatherings, which always happened on the weekends. A "farmers market" seems like an oxymoron, as the farmers I know don't like marketing and the marketers I know don't want to farm. But for many small farmers, the markets are also a chance to get off the farm and socialize a bit. Our social needs were not met in the marketplace. Rather, they came about through what I call farm festivals.

— CHAPTER THREE —

Farm Festivals

S teiner began the first lecture by speaking about the beautiful atmosphere on the farm at Koberwitz, so excellent and exemplary, and expressing his thanks for spending festive days there. Again at the end of the course, he said, "What we have here been doing as a piece of real hard work, work which is tending to great and fruitful results for all humanity, has been given a truly festive setting by our presence here," adding, "All that has been done . . . has placed our work in the warm and welcome setting of a truly beautiful festival. Thus, with our Agriculture Conference we have also enjoyed a real farm festival." Festivals played an important role in the peasant life of his childhood, and have influenced mine, as well.

My dad, born at the turn of the twentieth century, had a strict religious upbringing in an educated Dutch-speaking farming community in Kansas. They were not known for their frivolity. Dad's nature was intellectual and introverted, so it was up to Mom to foster the festivals at my family's farm.

Mom's parents came to the United States from Lithuania, and she grew up speaking their language and living around that community. This small Baltic country was a last holdout for pagans resisting persistent and domineering religions and regimes. Under the guise of Catholicism, these folks worshipped, understood, and lived with Nature, and they also knew how to have fun. Mom was vivacious and always up for merriment and shenanigans, as she called it.

Steiner knew how to have fun, too. Although photos of Steiner never show him smiling, it's said that he was quite a humorous man, and a bit of a jokester. Steiner mentioned old peasant calendars several times during the agriculture course, and he began the lectures with a quip from one: "If the cock crows on the dunghill, it'll rain—or it'll stay still." Try as I might, I can't find any deeper meaning in this saying other than—it doesn't really matter. He called this the needful dose of humor. Other little jokes were interspersed throughout the lectures, such as a remark that we have little opportunity to value the diamond, for we can no longer afford to buy it! He also said that university teachers who develop brilliant theories with loosely worded phrases suffer from the disease "Psychopathia Professoralis." You should have seen me trying to follow his advice that in America it might prove necessary to hammer the manure in the horn to make it denser. But it just squirted back out of the horn into my face. A horn can hold only so much manure, and I finally got the joke when reading in Steiner's lecture on geographic medicine that he thought the people in the Western Hemisphere were denser than those from the East. Steiner designed a famous sculpture called *The Representative of Humanity.* It shows a human figure posed between a devil above and a devil below, with another face in the upper corner of the sculpture. This face Steiner called humor, emphasizing its importance for humans.

* * *

As suburbs arose around our 40-acre homestead, it became the natural place for community gatherings. Mom's playful spirit balanced Dad's serious side, and we enjoyed many farm festivals such as school events, ball games, taffy pulls, and holiday celebrations along with birthday parties, bonfires, and campouts. Dad kept the gardens and the landscaping meticulously hoed and mowed, partly because he loved plants, and partly to showcase the beautiful plantings to potential customers.

Debby and I began hosting festivals soon after moving to Tennessee. At twenty years old and freshly out of the psychedelic sixties, we loved to party and play music with our friends. But at the same time, our organic homestead was working toward becoming self-sufficient. I knew then, as I do now, of the value of organic farming and renewable energy for reversing the negative effects of chemical agriculture, unlimited fossil fuel use, and overconsumption in general, and our farm festivals also exposed folks to nature and an alternative lifestyle. But mostly we just wanted to have a good time.

The festivals became extremely popular, attracting homesteaders, hippies, and rednecks throughout middle Tennessee, southern Kentucky, and beyond. I mowed the field and built a bonfire, and the only rule was to hold hands in a circle while it was being lit. This tradition continues almost fifty years later, and the collective energy of love and joy inspires and uplifts our community. These events were free and family oriented, held at the end of a long, winding road that crossed the creek seven times, a mile from any signs of civilization. We felt free, enjoyed wonderful potluck meals, and watched the kids grow from year to year. It was our own mini-Woodstock, with no rock stars, just us jamming with acoustic instruments and beating on conga drums until dawn. The summer solstice, the longest day of the year, was also the longest night of the year for many.

An unexpected result was the networking, long-term friendships, and community building that ensued. Farming and homesteading can be lonely occupations, and we were so glad to meet each other and make these lasting connections. For my part, I don't know if I would have wanted to live so remotely, far from electricity, phones, and running water, without these occasional, festive interactions with others from elsewhere. As I integrated into the local community, I learned about their quilting bees, hog killings, barn raisings, revivals, and other local get-togethers. I soon began hosting agriculture conferences to draw more attention to the importance of farming organically. These became "parties with a purpose," as one friend put it. Sometimes it seemed like the social aspects of farm life dwarfed the hands-on farming part, but over time I began to understand that both were integral to the farm community as a whole.

Farms are a great place to have parties and play music. We bought a circus tent for shade and shelter, and shared it with friends who wanted to host events in their fields. We made trails for walking in the woods, enjoyed the swimming hole unabashedly, meditated up on the hill, and rolled around in the hay. The CSA deliveries on Mondays kept our weekends free for festivals and fun. By 1990 our farm was hosting weekly volleyball games, including the typical potluck meal, music, and bonfire toward the end of the day. Simply questioning the myth of private property allowed me to open the land to any and all people. It seemed that within our farm's boundaries, magic happened.

When people ask me how to build community, my simple answer is—don't move. I grew up under very different circumstances than my neighbors, up north in a more liberal and fast-moving environment. After fifty years in Tennessee, I find I have absorbed some of the qualities of the folks around here, and that the people I interact with have absorbed some of mine. I learned to speak more slowly and developed respect for goodhearted but extremely conservative farmers, and I suppose they've gradually become receptive to me and my ideas. Communes came and went, but the community where I live remains relatively stable. It is an unintentional community.

One thing my neighbors and I have in common is that we don't want to move. Community continuity over decades fosters long-term thinking. Culture is created by people working, playing, and living together for a long time. Bringing our worn-out farmland back to life has taken hundreds of helping hands volunteering their time and energy. Any profit I make from farming is immediately reinvested into the infrastructure we are continually building and rebuilding on the farm. For us, the only profit in agriculture is the culture.

* * *

Throughout history, festivals celebrated the solstices and equinoxes, the annual agricultural turning points. As our farm festivals evolved, I began to inwardly feel the change of the seasons in our work at these times.

From a deep winter hibernation I emerge with boxes of seeds, plenty of plans, and a vision of healthy gardens, happy cows, and a few new projects. The spring equinox campout can be cold, but friends and family still come to enjoy nature's awakening. If I don't have peas, onions, and potatoes planted by then, I will very shortly. My mind and energy are focused on the future farming season, finishing tree grafting and berry propagation, spreading compost and plowing fields, and sowing the cold frames. The bare tree branches look like nerve endings tipped with pastel colors, hinting of the bursting buds and blooms to come. Appreciating foolishness, we always celebrate April Fools' Day.

Slowly the pastures are dotted with newborn calves and glow with new grass growth. As the ground warms up, I'm on autopilot, with no time to think, just following that instinctive urge to get it all planted. Spring, our busiest time of the year, is like a coiled spring waiting to be sprung. Everything planted needs hoeing before planting the next field. Other fields still need to be mown, composted, and plowed, and it's the time to make new compost piles. We make and use some of the biodynamic preparations at this time of year (more on that in chapter 8). By May, it is full tilt boogie with all the plowing, planting, and cultivating requiring constant attention. Vegetable deliveries soon begin. After hosting a preparation-making workshop on Memorial Day weekend, June arrives quickly.

Then a funny thing happens. While a small village begins to appear in a freshly mown hayfield, the sun peaks at the longest day of the year and different dynamics appear in the gardens. Weeds lose some of their persistent vigor, our crops begin to dominate the gardens, and we shift into weekly harvest mode. On the summer solstice weekend, the village in the field explodes in the exuberance of friends gathering together on St. John's Day around a huge bonfire. There's a stage with great bands playing, along with small improvised jams in the campground. We have made it through spring, and hundreds of friends help the farm celebrate. I often grab a few festivalgoers to help harvest garlic or mulch and stake the tomato field.

As our hoeing chores lessen, we move on to digging onions, then potatoes, and then gathering winter squash and watermelons. We are continually re-planting green beans, cucumbers, and summer squash for later harvests, and preparing ground for fall crops. Much of our food preservation happens during August.

Ripening fruits fill the late-summer days as we ease on into autumn. We raise apples, pears, and cherries, along with blueberries, raspberries, and grapes. Most years some of them make good crops, depending on late frosts, rains, and droughts. Two festivals loom ahead—the fall equinox and, a few weeks later, the biodynamic family reunion. The former is a rerun of the summer solstice celebration; the latter is an agricultural conference with lectures, dinners, and the making of biodynamic preparations. Both occur near Michaelmas (September 29), and our souls return from the hectic summer's long busy days to the steadier pace of fall farmwork.

Colors brighten the trees while we finish up fall haying and getting the ground ready for winter crops. The harvest continues, and the fall crops require tending, but we can feel the tension unwinding as we and the farm prepare for rest. I love to spread compost in the autumn, to help what is underground wake up again to digest the vegetable refuse from gardens, pastures, and meadows.

Sandor Katz leads a fall fermenting foods festival at the farm around the time of All Saints' Day, and folks learn about microbes in the soil and in their bellies while we chop and salt cabbages. Soon, the Thanksgiving festival shows off the summer's work. Then come the deep dark days of winter, but not before more festive times. The winter solstice gatherings at the farm are relatively small. We are outside and it can be chilly. A fire warms us, and maybe a batch of holiday cookies brings some cheer, but we are aware of what awaits, so we toast Christmas and New Year's and slip back inside under winter covers.

* * *

Festivals and religious holidays were important in Steiner's time, and he spoke about how the cycles of the year are mirrored in our inner soul life. His *Calendar of the Soul* creates a yearly picture of this transformation.* In it, fifty-two weekly meditations describe how our souls slowly open around Easter to receive what spring brings, become completely overwhelmed by summer's sense impressions, gradually return inwardly through the fall, and then sink deep into our selves during winter.

Steiner clearly delineated times of the year with certain inner and outer activities. For example, when addressing whether parasites could be combatted by means of mental concentration, he thought timing was important. You would want to choose the proper season, from the middle of January to the middle of February, to establish a kind of festival time and practice certain concentrations then. Inner quiet time in January and February is important for me, and one of the reasons I never wanted a plastic hoop house for winter production. I enjoy having some time off to plan gardens and pastures, picturing them in my mind. I clearly imagine crops growing happily in the fields, and healthy animals on lush pastures. I spend this "festival time" by myself doing the least amount of physical labor of the year.

At our community farm, a steady stream of visitors comes and goes. If an interest in biodynamics is expressed, we will stir a preparation. For many, *doing*

---

* I like the John Gardener translation.

teaches more effectively than *talking*. Speaking of stirring the preparations, Steiner imagined you would have many guests over and do it on Sundays as an after-dinner entertainment.

It was in 1987 when Harvey Lisle, Hugh Lovel, and I first decided to host a biodynamics festival. We felt it was important for folks to eat what we grew for three days to notice the difference in how biodynamic food makes them feel. This is what really convinces people, especially chefs and foodies, that biodynamics works. Although we have had the best of speakers, workshops, and farm tours at our conferences, the quality of the meals and the warmth of the participants are what bring people back. By hosting festivals annually at the same time each year, we have made them part of the cycle, like an annual family reunion. We feel part of each other's lives, as we experience the children, and ourselves, growing from year to year.

An insight for a traveling festival came to me after a visit to Hugh Courtney, our preparation-making mentor. His cellar held more preparations than he knew what to do with. They were inexpensive to make, so why not offer them to organic farmers for free? These folks would already have compost piles in the works and fields that could be treated—they just hadn't experienced biodynamics yet. There would be no need to include a lecture on biodynamics or to bother the farmers in any way. We tried out the idea, and it was a blast. We now call ourselves the Merry Prepstirs, and we've traveled in several parts of the United States, especially around biodynamics conferences. We are an ever-changing crew of elders who want to share our joy of working with nature this way. I envision that someday we'll have a bus equipped with folding tables, chairs, and our food, musical instruments, and plenty of preparations. We will visit any farmer who'll let us apply the preparations and camp out, and then take off for another farm.

Our farm festivals work well because we don't spend much money and rely on volunteers instead. The helpers become personally invested in the project, and we build a community while planning and preparing for the party. As we've seen, "festival" can mean one of many different types of gatherings. Because all farms are unique, so are their gatherings, if they have any.

My 1937 *Webster's Unabridged Dictionary* cites this derivation of the word "farm": "ME, ferme, rent, revenue, service; AS, feorm, provision, food, a feast, from LL. Firma, a feast; so called from the fact that the lands were let on condition that the tenant supply the lord with so many nights entertainment." In the old days, farmers supplied the lords and nobles with food and entertainment in exchange for protection. To farm meant to collect taxes. The "farmers" didn't do the work, they "farmed" it out to the peasants, who did what we now call farming, and then the "farmer" took a portion of the crops to the king.

I'm reminded of the question about who gets the very best strawberries from the king's garden. You might think it's the king. But the person who *picks* the berries gets the very best ones, the person taking them to the kitchen gets the next best, and then there are the cooks' helpers, the cooks, and the servants who present the king with, well, whatever is left. Our kitchen table gets the very best of what our farm produces.

Dad earned two doctoral degrees at Peabody College for Teachers in Nashville (now the education school of Vanderbilt University), and had careers as both a math professor and an industrial psychologist. His stories about friendly Tennessee influenced our decision to move there. By the time of his second family, which included me, he had already put back on his overalls and lived on a farm. So I learned early on to respect not only higher education, but also the hands-in-the-dirt knowledge that only comes from practical farming. Instead of a formal education, I decided to let nature and fellow farmers educate me. A different kind of intelligence comes from farmers, and they're the ones who I liked to hear talk about farming.

— CHAPTER FOUR —

Who Should Talk About Agriculture

E arly in the course, Steiner talked about university economists writing and lecturing on how agriculture should be carried out in the light of social and economic ideas. This irked him.

> One cannot speak of Agriculture, not even of the social forms it should assume, unless one first possesses as a foundation a practical acquaintance with the farming job itself. That is to say, unless one really knows what it means to grow mangolds [beets], potatoes and corn! . . .
>
> No one can judge of Agriculture who does not derive his judgment from field and forest and the breeding of cattle. All talk of Economics which is not derived from the job itself should really cease.

An economic idea from early-twentieth-century farmers called "parity" means that crop prices should reflect the cost of production plus a living wage, like the pricing structure that any business needs in order to survive for the long term. One of the ideals of Jeffersonian democracy was that a nation of healthy small farms was integral for democracy to thrive. But the land-grant universities, since the early 1900s, promote agribusiness controlled by non-farmers who are unconcerned about the long-term social, economic, and ecological effects of their advice. The non-farming experts from universities and corporations think they can talk about farming from a theoretical point of view because they look at a beetroot as a commodity separate from the farm. As non-farmers, they don't understand how a beetroot lives together with the soil and the field. The true cost of a beet is the totality of what it will take to grow it again next year—in other words, the cost of supporting the whole management of the farm.

* * *

After ten years of gardening in Tennessee (that would have been about 1983), I loved to talk about and show off our 1 acre of crops. I was really smart back then and no one could tell me anything. I helped form a Tennessee organic group and spoke at our conferences. Ten years later (1993), I was growing and marketing crops of beets, potatoes, and corn, among other things, but I wasn't quite as smart. I still loved talking about how I tended our 2 acres of vegetables, but my lack of understanding of exactly how plants grow was dawning on me. Nonetheless, I was writing a weekly newspaper column about gardening and farming and was soon to be the host of a television segment for the PBS show *Volunteer Gardener*.

Within the next ten years I was using a tractor with a front-end loader and a Farmall 140 tractor, which is small, lightweight, and has the engine offset for ease of cultivation. Two acres of vegetables had become eight by 2003. I was finally beginning to understand how the crops of beets, potatoes, and corn came about. The old-time knowledge I had gained from my family, neighbors, and old books was making sense, and the ideas I had read about in the *Agriculture Course* were slowly sinking in. I loved listening to farmers talk. Although my neighbors weren't organic, 90 percent of my work was just like theirs—fixing tractors, feeding cattle, and making hay. We loved our land, work, and community, and we probably had more in common than not. Thick-headed as we were, a mutual tolerance and respect grew between us. My neighbors began asking me for gardening advice, while I learned more about tractors, cattle, and other farm equipment from them.

In 2013, with forty years of farming under my belt, I felt I could finally speak knowledgeably about agriculture. I had learned to consider the bigger picture of how all the aspects fit together to create a whole farm organism. Now another ten years has brought me to the present, and I still feel I have so much more to learn. Steiner's quote humbles me. I made the mistake of talking about agriculture long before I really knew how to grow beets, potatoes, and corn.

* * *

For my first ten years as a farmer, I focused on compost. The soil on the Illinois farm where I grew up was the result of a glacial deposit within the last fifteen thousand years, and was very mineral-rich and full of humus. It took me years to realize the need to re-mineralize the soil on our Tennessee farm. The Southeast has not benefited from a glacier passing through and grinding up rocks at any point in the past million years.

During the next ten years, frequent applications of mineral-rich rock dusts brought notable results. Besides lime, I used rock phosphate, granite meal, greensand, and small amounts of a few others. But it wasn't until I spent a few days at a workshop with Australian biodynamic farmer Alex Podolinsky in 1992 that the concept of wise tillage took root, and I came to a greater understanding that soil is not static, it's alive.

Two soil scenarios were presented to the twenty attendees of Alex's workshop in Pennsylvania. One was in the beautiful garden beds with finely tilled soil that looked like compost, and the other in an old fence line in the pasture. The former was too fine and lacked structure, integrity, and crumb. When I dug up a shovelful of soil from a spot along the fence line, where no tillage had happened in a long time, we saw what he meant. The soil looked like cake—with pores of various sizes distributed throughout. (This type of structure is called soil "crumb.") This was the truly good soil, Alex explained, rather than the overly fine garden soil. This crumbly soil would remain structured after a downpour. The garden soil, being too fine, would compact and require more tillage, further destroying the precious structure and injuring microbial life.

The following spring I took Alex's advice when I planted a field that I previously would have wanted to till a lot more. The grass did re-sprout around the potatoes, but subsequent cultivation took care of the sprouting sod. After a 3-inch downpour, the lesson sank in, just like the rain did. When the soil dried, it did not become compacted. Instead, it sprang back loose and fluffy. This was the first year that Colorado potato beetles were not a problem on the farm. I believe now that beetles attack crops when they are growing in compacted soil. I remembered that an old-time farmer had also told me not to overwork the soil, just rough plow and gently harrow. It is hard to break the habit of tilling the soil too much, because of being anxious to plant. In farming, timing is everything, including knowing when to stop and wait. Doing nothing is often harder than doing something.

* * *

As a boy, Steiner planted potatoes, helped breed pigs, and lent a hand with the cattle on local farms. In the address after the third lecture, Steiner reminisced on his peasant upbringing.

> These things were absolutely near my life for a long time; I took part in them most actively. Thus I am at any rate lovingly devoted to farming. . . .

43

*As I look back on my own life, I must
say that the most valuable farmer is not
the large farmer, but the small peasant
farmer who himself as a little boy worked
on the farm. . . . In my life this will serve
me far more than anything I have subse-
quently undertaken.*

*Therefore, I beg you to regard me as the
small peasant farmer who has conceived a
real love for farming; one who remembers
his small peasant farm and who thereby,
perhaps, can understand what lives in the
peasantry, in the farmers and yeoman of
our agricultural life. . . .*

*It will always be a beautiful memory to
me if this Course becomes the starting point
for carrying some of the real and genuine
"peasant wit" into the methods of science.*

Living among farmers is the best way to understand
them, and to understand agriculture as a whole. These
are the people who should talk about farming, those
with their hands in the soil. Keeping that soil full of
humus is their concern, not economic theory. Only
farmers know the limitations on production necessary
to maintain humus, such as when to grow cover crops
instead of cash crops, or when there are too many cattle
on the land. You have to have your hands in the soil for
a long time before you understand how much to sell,
and how much to put back, or the relative amounts
of acreage of forest, crops, pasture, and hay that will

add up to a thriving farm ecosystem. There is a great difference between what you can learn from a farmer speaking of these things and what you can learn from someone who isn't a farmer.

Both farmers and scientists attended the agriculture course, and during that time they decided to work together. Each had something to offer the other. Throughout his lectures, Steiner spoke of melding instinctual wisdom with science as a way to understand agriculture as a whole. After all, scientists have been studying how plants grow for centuries, and their observations and research certainly offer farmers many valuable insights. Here is my summary of what just a few representative scientists, out of hundreds, have contributed to the scientific knowledge of how plants grow.

* * *

I learned the Latin names of trees from dad's nursery business and continued learning the botanical names of plants as classified by Carl Linnaeus, an eighteenth-century Swedish botanist. From orders and families to genera and species, it is fun to learn who is related to who. Linnaeus classified plants based on the structure of their flowers. For example, all the flowers in the lily family have three petals, the crucifers (cabbage family) have four, and the rose family has five. Flowers in the legume family look like little animal faces.

The German poet and scientist Johann Wolfgang von Goethe also studied plants in the eighteenth century. He looked for the commonality in all plants and wrote about it in *The Metamorphosis of Plants*, an easy-to-read book I fell in love with. The life cycle of sprouting, growing, and forming a bud to flowering, pollination, and fruiting offers an observable account of an archetypal plant.

Justus von Liebig must again be mentioned here too, because his research on the elements plants require for growth led to the development of the fertilizer industry. Later in life, though, Liebig changed his mind about the necessity for water-soluble nutrients. He realized the importance of humus and that it could release nutrients without them being soluble.

George Washington Carver, a professor at Tuskegee University from the late 1890s well into the twentieth century, was an early proponent of using organic wastes in farming, and promoted legume cover crops to restore fertility to worn-out cotton fields. He helped farmers by developing alternative crops for the South, such as field peas, peanuts, and sweet potatoes.

Agronomist Cyril Hopkins, who was on the faculty of the University of Illinois from 1900 to 1919, taught that phosphorus was the only element that might need to be added to soil. The USDA was teaching methods of self-sufficient agriculture back then, as this 1910 quote from Hopkins makes clear.

> *To maintain or increase the amount of phosphorus in the soil makes possible the growth of clover (or other legumes) and the consequent addition of nitrogen from the inexhaustible supply in the air; and,*

with the addition of decaying organic matter in the residues of clover and other crops and in manure made in large part from clover hay and pasture and from the larger crops of corn and other grains which clover helps to produce, comes the possibility of liberating from the immense supplies in the soil sufficient potassium, magnesium, and other essential abundant elements, supplemented by the amounts returned in manure and crop residues, for the production of large crops at least for thousands of years.

I believe he is describing the farming practices Steiner refers to when we are implored to go on manuring as before.

I'm a fan of the works by Peter Henderson, Liberty Hyde Bailey, Luther Burbank, and many others from this era. I also have studied and found much practical advice in my collection of farming textbooks and yearbooks of agriculture, many dating back more than one hundred years, that explain how farms were managed before the turn of the twentieth century. I condensed my studies of these old articles and books in chapter 10, "Go on Manuring as Before."

As discussed in chapter 1, the discovery of the synthesis of nitrogen at the beginning of the twentieth century led to fundamental changes in farming practices. By the early 1920s, Steiner was speaking out about artificial fertilizer's detrimental effect on soil humus, saying, "Mineral manuring [artificial fertilizer] is a thing that must cease altogether in time, for the effect of every kind of mineral manure, after a time, is that the products growing on the fields thus treated lose their nutritive value. It is an absolutely general law."

It became clear that we cannot replace by artificial means the fertility the earth itself is able to achieve by natural humus formation.

A decade later, William A. Albrecht at the University of Missouri published many papers detailing the importance of humus not only for plant health, but also for the health of the animals and the humans who ate them. A plethora of seminal works followed, including those by F. H. King on farming practices in China and Sir Albert Howard on those in India. The organic farming movement continually pointed to the millions of microbes in a bucketful of humus-rich soil as being the key to soil, plant, animal, and human health.

* * *

In the mid-1970s, Mom bought Dad four books she found in *The Whole Earth Catalog*, as he was an avid organic gardener. They were *Companion Plants* by Helen Philbrick and Richard Gregg, *Weeds and What They Tell* by Ehrenfried Pfeiffer, and Steiner's *Nine Lectures on Bees* and *Agriculture*. These influenced me, and I soon found three mentors in biodynamics. Harvey Lisle, a retired soil scientist, became a mentor not only in soil chemistry, but in trying to understand living forces as well. Hugh Courtney mentored me in the making of preparations. Hugh Lovel and I became fast friends, reveling in our mutual mistakes, blunders, and occasional insights.

Now that I have been farming for a long time, I feel the three pillars of agriculture are biology, minerals, and tillage. Simply reading about the pillars did not bring me understanding. We don't learn instinct from books. It took decades of growing beets, potatoes, and corn to really possess the foundation of the farming job itself. I had to find the peasant in myself to balance out all that I had read.

After fifteen years of stirring small amounts of biodynamic humus products in water, I learned about another way to do this called compost tea. This is made by putting compost in water and using an aquarium pump to aerate it for a day. The purpose of the aeration, either by vigorous stirring or by pump, is to add oxygen so that the beneficial microbes in the compost can propagate rapidly. The great thing about compost is the abundance of beneficial microorganisms, and these methods are a way to increase their population in the soil where they are applied. Scientists working with ever more powerful microscopes were finding not millions but *trillions* of microorganisms in a bucketful of soil. To illustrate the difference in these numbers, imagine a stack of hundred-dollar bills about 3 feet tall. That's a million dollars. Now picture a stack 600 *miles* tall. That's a trillion dollars. Counting microbes is like counting stars—the stronger the lenses, the more you see. Pictures of stars from the biggest telescopes are eerily similar to photos of fungi from the strongest microscopes.

Trying not to let science get in the way of learning directly from nature, what I really like is blending the two. I also try to balance organic materials with minerals, tillage with cover crops, grazing with rest, and work with play. As I integrated my practical experiences with neighbors' advice and a bit of book learning, the farm thrived. I plowed deeply but carefully, subsoiled along contours when dry, remineralized with rock dusts, incorporated crop residues and decomposed forest products, and grew as many cover crops as I did cash crops with as little tillage as possible. I moved forty head of cattle around, fed them the hay I baled, and made hundreds of tons of compost from their manure and farm wastes, which a biological assay rated highly. I applied that compost liberally to our crops and they flourished.

It appeared that the more organized the living entities were, the more energy they could gather. Life begets more life. Although material things displayed entropy, the running down of energy, living things worked together to accumulate energy. This is called syntropy. My machines broke down, but the gardens and herd built up.

Still wondering about how plants grow, I kept thinking about new scientific insights regarding fungal hyphae reaching far and wide into the soil in search of nutrients for their host plants. My practical side noticed that the better the quality of the compost, the better the crops. There is an intelligent design in nature, and this was evident in healthy soil growing healthy plants. Steiner seemed to be talking about fungal hyphae in his fourth lecture.

It is simply untrue that the life ceases with the contours—with the outer periphery of the plant. The actual life is continued, especially from the roots of the plant, into the surrounding soil. For many plants there is absolutely no hard and fast line between the life within the plant and the life of the surrounding soil in which it is living.

The sugars formed through the daytime activity of photosynthesis sink down and out through the roots as tension in the plant relaxes at night. These feed the soil life specific to that plant species. This life fixes nitrogen and releases minerals in the nighttime, and sap pressure brings them up into the plant the next day. We've come back around to that main insight from the *Agriculture Course* that nitrogen is not an agricultural input. Rather, nitrogen is the *result* of good agriculture. Steiner can help us learn about nitrogen and her "four sisters."

CHAPTER FIVE

Nitrogen and Her Sisters

S teiner sure makes chemistry interesting. As he describes the basic elements the farm needs to grow crops and animals, they come alive with individual personalities. Life, through plant growth, seems to be a matter of the "girls" in the air joining with the "boys" in the soil. Steiner was a scientist who also had a profound understanding of the history of scientific thought, philosophy, and religion. Along with his own personal history of growing up among peasants and studying the way traditional country people thought, he was in a unique position and time to talk about agriculture and humanity's relationship to it.

The "girls" are nitrogen and her "four sisters." Steiner introduces them in the third lecture. "The four sisters of nitrogen are those that are united with her in plant and animal protein. . . . I mean the four sisters, carbon, oxygen, hydrogen and sulphur."

These five elements are the free elements or the atmospheric influences, and together they make up 95 to 98 percent of the substance of a plant. This helps explain why air in the soil is so important. If the soil becomes compacted, I have to loosen it up somehow. Tillage is one way, but then I must reinvigorate the microbial damage that tillage can cause by spreading compost and managing a system of diverse cover crops and rotations. Keeping the soil covered in plants really helps, as does mulching, as they mitigate the effects of sunlight baking the ground and heavy rainstorms beating on it. In our clay-based soils, rain on bare ground mixes with small clay particles and "runs together," another way of saying that the soil structure collapses and turns to mud. When the mud dries, the soil hardens up. Growing grass for a few years helps rebuild the soil structure.

Steiner advised farmers to learn to know the "deeper essence" of these five elements with some words about carbon. "When the old alchemists and such people spoke of the Stone of the Wise, they meant carbon—in the various modifications in which it occurs," Steiner said. "Carbon, in effect, is the bearer of all the creatively formative processes in Nature. . . . It bears within it the creative and formative cosmic pictures—the sublime cosmic Imaginations, out of which all that is formed in Nature must ultimately proceed." He is describing nothing less than how life incarnates. Because it has four available electrons that can make so many different bonds, carbon is the versatile, life-creating element.

The element that carries the influence of life into carbon is oxygen. Oxygen bonds with the elements of the earth, creating compounds that can be taken up by roots so they can be involved in the various life processes in the growth of plants and animals on the farm. Oxygen bonds with hydrogen to form water, with potassium to form potash, with phosphorus to form phosphate, with sulfur to form sulfate, with nitrogen to form nitrate, and with carbon to form carbon dioxide. Anything that was once living contains carbon, and while alive it includes oxygen. For plant growth, carbon and oxygen need another element to bring them together.

Steiner assigned this role to nitrogen, stating, "Everywhere—in the animal kingdom and in the plant and even in the Earth—the bridge between carbon and oxygen is built by nitrogen. . . . The nitrogen has an immense power of attraction for the carbon-framework." Nitrogen mediates between carbon and oxygen. As air enters the soil, it becomes part of the living beings there. While carbon supplies the form and oxygen the life, it is nitrogen that becomes sensitive inside the earth. This sensitivity is of immense importance for agriculture. Nitrogen is everywhere present in the air, and thus, in Steiner's view, nitrogen knows what's going on everywhere on a farm. It has a sympathetic feeling when the proper amount of water is present or the right kinds of plants are growing on the farm, and has a feeling of antipathy when things are out of balance. Nitrogen and carbon (along with hydrogen and oxygen from water) come together to make amino acids, which are the backbone for proteins, DNA, and everything that makes life function.

Sulfur also needs to be present, according to Steiner.

It is along the paths of sulphur that the
Spiritual works into the physical domain

*of nature. . . . Hence the ancient name,
"sulphur," which is closely akin to the
name "phosphorus." The name is due to
the fact that in olden time they recognised
in the out-spreading, sun-filled light, the
Spiritual itself as it spreads far and wide.
Therefore they named "light-bearers" these
substances—like sulphur and phosphorus—
which have to do with the working of light
into matter.*

Through the conventional lens of science, we understand sulfur through its action as a catalyst in chemical reactions, similar to how grease functions in gears. It is one of the trace elements that plants require in small amounts for many specific biological functions to happen. As a catalyst, though, the trace element doesn't get used up in the process. Sulfur is a catalyst for photosynthesis, which is powered by light; light into matter, through plants.

When I began farming, the soils of the Southeast had too much sulfur because of acid rain from burning coal. The coal power plants now have scrubbers that remove the sulfur, so now if the soil is deficient in that element, we add elemental sulfur, applied in small quantities because it's very acidifying. I spread half a cupful around each blueberry plant annually to keep their soil pH near 5.5, which is what they like. (Most crops prefer a soil pH closer to 7.) Gypsum, calcium sulfate, is used when you want to add calcium or sulfur but not alter the soil pH. Calcium is alkaline and raises the soil pH.

Steiner mentions phosphorus, a macronutrient in the soil, very little other than to describe it as a "light-bearer." It comes up in relation to tilling the soil and manuring properly, and also when describing the use of diluted valerian juice as a biodynamic preparation to stimulate manure. As much as Steiner discusses the other elements in his lectures, I find it odd that he gives phosphorus so little attention. Steiner aligns it with sulfur's role, apparently a very important one as the carrier of the spirit. Although phosphorus can be a limiting element in some farm soils, too much can be a problem because it will tie up other nutrients and make them unavailable.

The four sisters can only work together with another one. We must include hydrogen. Hydrogen establishes the connection for our physical world with the wide spaces of the universe. When the spirit no longer wants to be physical, hydrogen, moistening itself with sulfur, carries it back out into the universe. I guess you could say this is how life disincarnates. Hydrogen is by far the most abundant element in the universe, yet it is also by far the tiniest and lightest, big in context yet small in content. These five sisters form the structures for the spirit to become physical here on Earth.

When something incarnates on Earth, sulfur carries it out of hydrogen and into oxygen, where life is found. Then, nitrogen guides life into the form that is embodied in carbon. I picture hydrogen coming from sunlight, or the universe as a whole, into Earth's atmosphere and joining up with carbon dioxide through photosynthesis in plants to form sugars that flow down through the phloem and out of the roots. These exudates awaken and feed the soil microorganisms, which incorporate nitrogen into amino acids that go back up the xylem and feed the plant the next day. Sulfur greases the surfaces to keep this all flowing smoothly.

A major part of what I do as a farmer is aimed at enticing the inert nitrogen gas in the atmosphere to engage with the life processes in the soil. This is done by microbes transforming it into amino acids that can then form proteins in plants and animals. Another important task is to promote oxygen's bonding with minerals to make them available. In other words, I use compost and cover crops to promote humus formation, and I cultivate when the soil needs air to incorporate the five atmospheric elements that come to life underground. The microbes do the rest.

These five sisters have distinct personalities. Sulfur is the friendly flitting busybody, a social butterfly, staying on the surface and helping things move along. Carbon, the solid, stable sister, is a sculptor, an artist in the shaping of structures, an anchor for the community. Oxygen is the promiscuous one, joining up with many different elements, fun and flowing, the life of the party. Nitrogen is the aloof sister, sensitive, sympathetic, and intelligent, who is content by herself, a mediator who needs coaxing and enticing to get involved. Hydrogen is the sister closest to the spiritual but also the least spiritual. According to Steiner, she is either driving the other sisters into the chaos of the protein in a seed or sending them out into the chaos of the far reaches of the universe.

These "girls" don't work independently; they are bound to and dependent on the elements found in the soil. Although Steiner did not call them brothers, I think of the substances of the earth—calcium, silicon, potassium, magnesium, sodium, and the trace elements—as the boys.

Carbon needs help in forming the physical structures of plants and animals. This helper role is fulfilled by calcium and silicon in the form of limestone and silica. Nitrogen carries oxygen into the soil to unite with carbon there. Now the carbon is able to join with silica and limestone and form the structures of plants. This leads into a discussion of legumes, and the idea that the limestone in the earth is dependent on a kind of nitrogen in-breathing, similar to the human lung's dependence on the in-breathing of oxygen. Steiner said that this is the kind of vision we must unfold when we look out over the surface of the Earth, covered with plants and having beneath it the limestone and the silica, the calcium and silicon.

So many different processes happen down in the earth, where the boys hang out, that scientists describe it as the soil food web. The more aerated and humus-rich the soil is, the more easily the elements in the soil air (the atmospheric influences) can interact with the nutrients locked up in the earth and free them to grow our crops. All that is working above the earth in the atmosphere is in mutual interplay with what is working underneath the ground, in the earth. The farmer's job is to facilitate these interactions, to help nitrogen and her four sisters join up with the boys in the earth.

— CHAPTER SIX —

Earth Elements

S ilicon and boron play different roles in plant growth than the other earth elements. Silicon provides the biological transportation system in plants. It's found as part of the walls of the tubular structures in a plant's vascular system, the phloem and xylem, and on the surface of the fungal hyphae that connect to roots and stretch out into the soil. Together, the vascular bundles and the fungal hyphae create a path through which sugars flow into the soil and nutrients move up into the plants. The insolubility of silica holds the nutrients inside the tubes as they move into the plant by boron-powered sap pressure during the day, and are released as exudates into the soil at night.

Any attempt to simplify chemistry or physics (or Steiner for that matter) quickly leads to mistruths because it is simply so complicated. But I'll try my hand at explaining how boron facilitates sap pressure by bonding with silica. In atomic theory (which is only a theory), a number of shells containing electrons surround the nucleus of the atom. The outermost shell often has missing electrons, leaving spaces that can be filled by electrons shared with other atoms. Atoms form bonds by sharing electrons to complete their outer shells and become stable. A silicon atom has four spaces and an oxygen atom has two, so silicon needs to share with two oxygen atoms, making silica (SiO_2). Boron's outer shell has only three electrons, so bonding with silicon leaves it "thirsty" for one more electron. This electron is supplied by hydrogen, leaving a positively charged hydrogen ion and thus a slightly positive charge on the inner walls of the tubes, creating sap pressure.

Sometimes farmers and gardeners need to add boron and silica to the soil. We find boron in clay, which is a good reason to add clayey soil to compost piles. Getting boron to entice insoluble silica to engage as the connector between the plants and the minerals of the soil requires careful farming. Farmers can add boron to soil in the form of borax, but it should be used sparingly. Very small amounts suffice. Silica is in many of the rock dusts we use.

The importance of silicon and boron cannot be overemphasized. They can easily become deficient in two ways that are common today. One is by the use of artificial fertilizer or organically acceptable fertilizer products that contain nitrate salts, including guano, fish emulsion, feather meal, cottonseed or soybean meal, raw manures, or poultry litter. Always look at the label. The other practice that interferes with silicon and boron is tilling too much, too fast, or under the wrong conditions, which can happen on any farm.

Half of the Earth's crust consists of silica, the oxide of silicon. Silica is found on the outside of living things, in the skin, in cell walls, and basically in the outer layer that separates the inside of anything alive from the rest of the world outside of it. Steiner devoted a lot of time in his lectures to silica's role in agriculture, pointing out that it plays the greatest imaginable part in life. I had only known silica as an insoluble crystal that fractures into hexagonal pointed shapes, so Steiner's claim seemed odd to me at first. But when I allowed myself to imagine silica as the interface between living organisms and the rest of the world, it began to seem immensely important. Not unlike nitrogen, silica requires life to engage it. It also needs warmth; silica can't work when it's cold outside.

Although there is plenty of silicon in the earth and nitrogen in the air, these most important elements for plant growth happen to be the hardest to break free from themselves. Thank goodness for lime, the oxide of calcium, which is the great engager. As the opposite of silica, which is found on the periphery of living things such as our skin or a peel, lime is found inside of living things, like in our bones and in seeds. In clay soil, calcium can get in between the tight layers of the potassium silicate in the clay and fluff it up. It draws down what is taking place in all the activity above the soil on the farm. Lime is very soluble, and we need to spread it often on our mineral-deficient soils at Long Hungry Creek Farm. Our soil tends to be acidic anyway, and in our humid climate, calcium continually leaches out of the topsoil.

Limestone is the fellow who would like to snatch everything for himself, Steiner claimed. Calcium has a greedy nature, constantly craving everything with no rest. Even after calcium combines with oxygen, it remains unsatisfied. So it bonds with other nutrients in the soil, mostly at night, and sap pressure carries them into the plant through the silicious tubes of fungal hyphae and vascular bundles during the day.

Steiner called silica the aristocratic nobleman who wants nothing for himself, but who can help carbon create the forms in nature. Lime's attraction to the other nutrients in the soil, along with its ability to move with sap pressure, can be compared to a vehicle. The silicious tubes are the road that lime travels with its load of plant nutrients. As farmers, our activities must support the two polar influences of silica and lime. Examples of these polarities are delineated in "Polarities of Silica and Lime" on page 63.

Silica works during the warmth of the day, helping plants with photosynthesis, blossoming, and fruiting. Its nature is vertical, like a mountain or the seed stalk of a plant. Lime processes happen at night, releasing the gathered minerals, helping microbes to release nitrogen, and generally digesting the daytime activities that sink into the soil as tension in the plant relaxes when the sun goes down. Lime's nature is horizontal, like the canopy of leaves stretching out laterally to catch as much sunlight as possible.

Fungal activity supports much of silica's work. This is why organic farmers refrain from too much tillage or the use of nitrate salts, both of which harm fungi and shut down silica's activity. We add rotted forest products to promote fungi. The ratio of fungi to bacteria in a forest is quite high compared to that in a meadow or a garden, where it is about even. Most

Polarities of Silica and Lime

Silica	Lime
Non-soluble	Very soluble
Igneous rocks	Sedimentary rocks
Crystal	Chalk
Silicon	Calcium
Outer planets	Inner planets
Mountains	Plains and hill country
Western United States	Eastern United States
Stems	Leaves
Sharp-pointed leaves (grasses)	Soft, round leaves (legumes)
Vertical	Horizontal
Summer	Winter
Works in daytime	Works in nighttime
Photosynthesis	Mineral release
Blossoming	Nitrogen fixation
Fruiting and ripening	Digestion
Food; nutrition	Growth; reproduction
Forms route for transport	Forms vehicle for transport
Promotes fungi	Promotes bacteria

crops appreciate leaf mold and other woodland humus because of the diversity of elements and nutrients fungi help make available. Because of its high carbon content, any kind of wood must be fully decomposed to prevent tying up nitrogen.

There is one family of plants on our farm that prefers soil that has become more bacterial through tillage. The brassicas thrive when we plant them in areas where cultivated crops of potatoes, beets, or corn have been grown. We plant brassicas only in the fall, so they mature in cooler weather. By not growing any in the spring, we avoid a lot of insect problems, such as cabbage loopers and harlequin bugs. Apart from brassicas, we usually sow a non-tilled cover crop after a cultivated crop to reinvigorate the soil's fungal activity.

Magnesium is an earth element that is found in plant cells, just as iron is present in blood cells of humans, and is essential for photosynthesis. Too much magnesium in soil makes other nutrients unavailable, and it will really tighten up the soil, making it sticky. It also causes the soil to hold moisture too long after getting wet and to become hard when it finally dries out. The remedy is to apply a high-calcium lime, because calcium opens up soil structure and promotes crumb formation.

The cation exchange is the measure of the percentages of the soil cations, the positively charged ions in the soil. If the magnesium were low, you would use dolomitic lime, which is high in magnesium. The calcium should be approximately 65 to 75 percent, magnesium 10 percent, potassium 5 percent, and sodium less than 1 percent of the soil's cation exchange capacity.

Potassium, a major element necessary for plants, is found in wood ashes, granite meal, greensand, and many other rock dusts. Soils contain large reserves of unavailable potassium, and biological activity can release it from these reserves. Ashes and rock dusts also contain trace elements, whose importance in very tiny amounts Steiner recognized long before others. Along with sulfur and boron, the other trace elements necessary for plant growth are iron, zinc, manganese, molybdenum, copper, chlorine, and cobalt. Researchers now have expanded this list to include selenium, nickel, aluminum, and iodine. Trace elements are also present in biodynamic preparations and in sea products like kelp and ocean minerals. (Note that the latter needs to have excess sodium removed before applying it to soil.) Adding potassium in the form of ashes and rock dusts to a humus-rich soil stimulates the biological cycle and keeps potash available.

Specific microorganisms, including actinomycetes, as well as fungi that help build humus, also need these trace elements so they can fulfill certain roles in biological processes. Trace elements are essential components of enzymes, hormones, and auxins.

Careful observation is a good way to spot areas where elements may be deficient. In good soil, deficiencies of these elements are indicated by characteristic signs. It's not easy to distinguish a plant disease from a nutrient deficiency. It requires constant observation to watch for progression of symptoms from day to day. Diseases spread fast. Deficiencies may appear suddenly, but the symptoms don't spread. Many plant problems that we think of as disease, such as blossom end rot on tomatoes, are actually caused by a mineral deficiency—in that case, calcium.

The presence of specific weeds can indicate the presence or absence of certain elements in the soil. Clover only grows well with lime and phosphorous available, while broom sedge tells us they are lacking. Thistles abound where there is excess potassium in the soil. As I mentioned, *Weeds and What They Tell* by Ehrenfried Pfeiffer was one of the booklets floating around in our house when I was young. I thought it was so cool to be able to know things about the soil simply by identifying the plants growing there.

* * *

Balancing the amount of animals with cropland, forests with meadows, and grazing with rest from grazing is a constant challenge. A farmer's experiences and instincts are helpful in this regard. My farm has suffered from too much tillage, but also not enough. I've allowed forests to encroach on meadows, later cutting out the cedar trees that I'd let grow up, and I've both over-grazed and under-grazed the pastures. I had to make these mistakes and observe the results, because that's how I learn. In his lectures, Steiner emphasized that in ancient times it was necessary to have an instinctual knowledge about entering into the inwardness of nature. Although this had to be eclipsed for the rise of intellectual thinking, he believed it was again necessary for science to recognize the value of instinctual, observational knowledge.

The silica/lime polarity is one concept that can help me decide what the farm needs. Microbes need calcium and air, but too much lime or tillage can hurt them. Although our crops need nitrogen, the wrong form of it impairs silica's activity. Tillage helps release nitrogen but hurts fungi. It seems everything I do both harms and helps. Goethe said that everything in nature lives by give and take. This sure is true on a farm. The cropping and grazing take, while the cover crops, compost, and organic wastes give. Our farm always has beautiful fields, and totally messed-up ones. The alteration of the two is like breathing. When asked about tillage, I say: no-till until you need to till, then till until you can no-till. The bottom line is to keep a sense of humor.

When making a small garden bed, I add sand if the soil is heavy clay. In a clayey field I grow a siliceous plant like a grass or grain crop that will finely subdivide the soil with its roots, as I can't add enough sand to make a difference. Potatoes grow well after such a cover crop.

Our sandier soils are in a flood-prone bottom, so we make hay there and feed it up on the hill where the cattle graze. I move nutrients around just like the animals do. Birds, reptiles, and mammals are all poking around the soil, moving minerals and organic matter around, and leaving valuable soil-building wastes behind. A good way to re-mineralize the farmland is to give minerals to the livestock, who will distribute them over the fields in their droppings. They know to lick the mineral salts that are missing from the farm's soil. Remember that everything in nature is interconnected. Even in a small garden, the earthworms and other soil life are moving nutrients around. This constant mixing up of the diverse components on the farm helps ensure that whatever is needed is within easy reach.

Potatoes don't want lime because it makes them susceptible to a disease called scab. So I add lime after harvesting them and then sow the fields in beans near the end of July. Beans love lime. Before the last cultivation of the beans, I toss out seeds of a cover crop mixture of buckwheat, crimson clover, brassicas, and maybe wheat, rye, peas, and vetch. The greater the diversity in our cover crop mixture, the more possibilities there are for enriching the soil with a wide range of nutrients and microbial activity.

Earthworms love the sulfur activity in the strong-smelling turnips and daikons and are often clinging to those roots when we pull them. I love earthworm slime. Its sparkling clarity reflects the rainbow and is colloidal in nature. Colloids can hold nutrients in an exchangeable form, but the nutrients are not water-soluble, so they don't leach out when it rains. Earthworms are soil regulators, ingesting lime and silica for grit and incorporating small amounts of those and other earth elements in a more available form into their castings. Plants love worm castings. Earthworms love lime but also free up silica for the soil microbes. Their eggs seem to be everywhere, and if food is available, they'll hatch and proliferate. Good soil has big, dark, shiny worms, whereas poor soil has small, light, dull earthworms.

When I hold a piece of limestone, I can feel it pulling moisture from my fingers like a piece of chalk would. The lime wants to interact. Holding a crystal feels different—no interaction. Along with silica, lime is also available in rock dusts. Since these compounds are not combustible, they are present in wood ash, too. Trees bring up minerals from deep soil levels that the roots of annual plants can't reach, so their ash is mineral-rich and valuable. We spread it lightly on the land, alternating fields from year to year with the crop rotations. Burning bamboo, which is a type of grass, will yield silica-rich ash that also contains potassium. The humus in compost, worm castings, and leaf mold helps make these minerals available for our crops. This is another reason that maintaining humus is essential for farming. Humus is the place where the elements from the air, the sisters, meet the boys in the soil.

CHAPTER SEVEN

Humus

Steiner repeatedly implores us to grow our crops in humus-rich soil. Humus is created from plant life being absorbed by the whole nature process, especially plant life that has not yet gone forward to the stage of seed formation. This is why I like to mow cover crops when they begin to flower. Once seeds form, the mineral elements are locked away in these storage vessels that make the new plant. Catching a crop at the right stage thrills me, as the white buckwheat, purple vetch, or very crimson clover are in peak color. Within a few hours the beautiful flowering field is laid flat, covered in compost, and gently chisel-plowed. So it goes.

The connections of the earth elements and air elements in the humus can be fostered when we fertilize with compost. Steiner was clear on its importance. "Humus and humus again should be given to the soil in every conceivable form—as compost, leaf-mould, etc." And in his fourth lecture, he stated the following.

> Manuring *and everything of the kind consists essentially in this, that a certain degree of* livingness *must be communicated to the*

soil, and yet not only livingness. For the pos-
sibility must also be given to bring about in
the soil what I indicated yesterday, namely
to enable the nitrogen *to spread out in the*
soil in such a way that with its help the life
is carried along certain lines of forces, as I
showed you.

This means that we must grow cover crops, especially legumes, and be extremely careful with our
tillage, which compromises the potential of fungal
hyphae to enable nitrogen to spread out in the soil.
After repeatedly arguing against the use of artificial
fertilizers, saying that they fertilize only the water in
the soil, Steiner recommended compost, which gives us
a way of kindling the life within the earth itself. From
my father's infatuation with compost to my neighbors'
insistence on fermented manure, and from my own
studies and practical experience, I believe compost is
the heart of farming. When lecturing I often have two
answers to every gardening question, either "add more
compost" or "I don't know." This is an oversimplifica-
tion, but I've seen good-quality compost pull off many
garden miracles.

Steiner has good composting advice. Don't pave
the site or dig a pit, just build a pile above ground level.
Air and rain are good for the process. Add enough carbon to prevent the nitrogen from evaporating into the
air as ammonia. Our noses are helpful in this regard.
I try to bring the compost heap into such a condition
that it smells as little as possible by piling manure (a

high-nitrogen material) up in thin layers, covering it layer by layer with something else, for instance old hay or chopped-up leaves (which are high in carbon), and then adding good garden soil before repeating the process. This uses carbon to attract and preserve nitrogen from escaping as ammonia, to be used later for growing crops.

I like to make compost piles where there's been one before. Some of the compost from a previous pile is always left behind, which seems to help the new one. I like leaf mold or humus from an old woodchip pile because the forest products contribute fungi. After forming the pile, I indent the top so that rain soaks in, but after a year or so I make the top pointed so that it won't become too soggy. A covering of hay or some other material is a good idea to prevent weed growth and to give the pile a skin.

If rock dusts are readily available, I add them— phosphate, greensand, granite meal, or basalt. Lime can be used sparingly, but too much hurts microbes and can cause nitrogen to volatilize. Lime is better used in heaps that don't contain much manure. Sometimes I also sprinkle ashes, borax, and elemental sulfur on a pile, being careful to use small amounts, as they can be detrimental to life in large doses.

I am not a fan of turning a compost heap often, but I do keep an eye on it. It shouldn't be allowed to get too hot, because enzymes and other life can be harmed at temperatures above 130°F. I have never had any pathogen problems in my compost that I know of, and my soil already contains so many weed seeds that hot composting is not going to stop weeds. If a pile does get too hot, I turn it to bring in air. If it dries out too much and a white mold called fire fang forms, it needs some water. Apart from that, I simply let a pile sit for six months and then I flip it, rather gently, to expose the insides to the influences of air and rain. After this initial breakdown of materials, humus formation begins. The pile can sit for a year or two now and the microbes will create a beautiful, black compost that will smear across your fingers like wax when you rub it in your hand. This texture and appearance indicate good fungal activity and that the compost is ready to be spread.

We use an old horse-drawn, ground-driven manure spreader. I have owned several and prefer New Idea models. My biggest one holds a ton and a half, so twenty loads is about thirty tons. I don't get hung up on application rates. A garden of about one acre gets about twenty loads. After spreading seven or eight loads, I can't remember if it is seven or eight. Farming is not rocket science, and after years, the precise amount a plot receives doesn't matter much. I just use what the farm has, putting more compost in places I feel need an extra boost. The results of liberal amounts of good-quality compost last for many years. The compost I make this winter may not be spread until years from now, but it will continue to help feed the soil for years after that.

I bought my first manure spreader in 1989 and was quite proud of my ability to load it seven times in one day with a pitchfork. It was my virgin experience using a machine other than tractors and hay equipment, but it didn't break immediately like I expected. By the time it did I was used to fixing hundred-year-old stuff. As we like to say, breaking farm equipment means you are farming. Within a year I was borrowing a neighbor's front-end loader and spreading dozens of loads in a day, and I attribute the subsequent mechanization and growth of the farm to that good manure spreader and my neighbor's generosity.

Speaking of manure, Steiner says that it has the force to overcome what is inorganic in the earthly element. I often picture microbes when Steiner says force, but his concept of "forces" is much broader than that. He makes the distinction that the microbes just indicate the goodness of the manure. Either way, it is compost, or fermented manure, that frees up the large reserves of minerals unavailable in our soils, which are the inorganic components. It is for the best that the large reserves of unavailable minerals are "locked up," as this keeps them from leaching out of the soil when it rains. As plants grow and release root exudates, dormant microbes wake up and begin to propagate. These microbes, living off what the plant releases, have a vested interest in the health of the plant. When they receive signals in these root exudates that the plant needs certain nutrients, the microbes are able to access the "locked up" nutrients to ensure the plant is fed, so they in turn will be fed later. You can see why organic gardeners and farmers love rotting garden refuse, cow manure, and compost.

Our vegetable leftovers, when not fed to a pig, are simply tossed out the kitchen door and recovered before too long and added to the piles. The rotting stuff that results from sorting vegetables, which on our farm is considerable, also ends up in the compost piles. Compost, my favorite thing, comes from some of the worst-smelling things, which I find to be an endearing characteristic.

* * *

I invited four types of farmers to speak on soil improvement from their particular perspectives at the annual Tennessee Alternative Growers Association conference in 1989. (We used the term "alternative" in the name because "organic" had negative connotations when we formed in 1982.) I was expecting to hear lectures on composting from the organic farmers, fertilizers from the conventional farmer, and esoteric preparations from the biodynamic viewpoint. Instead, they all said the same thing—grow grass and clover. In my early gardening experience, I considered grass and creeping clover as problems, getting in the way of my crops. Now I was being told they were a solution. Of course, the farmers were right. The grasses were trying to heal the soil structure I had destroyed with unwise tillage, and the clover wanted to open up deeper layers of the soil profile and bring in some air elements. I began to take cover cropping and pasture management more seriously.

Cattle eat grass and clover, so all of them together can be used to build humus by controlled grazing and long periods of rest for the pasture. Even when in a field of 10 acres, the herd will all mob together in a small spot.

They love to be close to one another. Graziers often use movable electric fence or set up small permanent paddocks so the herd mows and defecates on a small area and then is moved daily to a new paddock. This allows the grazed areas to completely recover before the cattle return. Grass forms a carbon-rich thatch that remains after grazing or mowing. Along with its roots, which die in proportion to the amount of tops removed, this thatch mixed with the wastes from cattle builds soil humus. This is carbon sequestration, the movement of carbon from the atmosphere into soils. Plants make soil, and do so particularly well after adequate composting, mineralizing, and animal impact.

Each plant species is associated with specific microbes and minerals. For example, clover hosts rhizobium bacteria and loves lime. Grasses work with fungi and silica. Clover and grass are companions, and together they are the great healers of the soil. Because tillage makes soil run together, after a while it really helps to put a tilled field in pasture or hay for a few years. Pastures get compacted from animals and may need plowing to loosen the soil and bring in air. Without compaction, clover brings in air and grasses keep the soil loose.

I was studying mob-grazing and holistic management when the thought occurred to me that the farm would be better off without cattle for most of the year. Instead of trying to manage the grazing of forty head, make hay, rotate pastures, and all that, I could have two thousand head for a day in May, and maybe again for a day in the fall. Think of what an impact they would have. Every plant would be tramped and defecated on, a total disaster. But with no more animal impact afterward, the pastures would grow back better than ever. I would simply have to be prepared for the onslaught, with my crops, home, and farm perimeter securely fenced off. This crazy notion made me realize that the idea of a self-contained farm individuality should include thousands of acres, as in Steiner's day. Nowadays it would require farm neighborhood cooperation and would make more sense in the Midwest where there is so much contiguous flat land.

Incorporating crop residues and cover crops is an excellent way to build humus into the soil. But it can be a pain in the neck, as the lusher it is the more stuff there is to incorporate. Chopping it up well really helps. Our farm, by design, uses the common equipment local farms use. First, I mow the field in low gear with a bush hog, usually two times. Even after that, though, the mowed material clogs up as I move slowly through the field with the chisel plow. So I lift the hydraulic arms and back up the tractor so that the stuff falls off. Then I pull forward, let down the hydraulic, and back up so the chisel plow digs into the soil from where I left off. Then I pull forward a few yards and get to do it all again.

Although this two-step-forward and one-step-back process, common in my life, is frustrating and time consuming, the reason I do this is because the more organic matter, the better. Slowly the crop residue or cover crop is more or less worked into the ground, and then I wait. Doing nothing is really hard for an excitable farmer. Instead of continuing to work the soil, I let it rest for a few days. But the soil is not resting. An underground battle is raging. Microbes feeding on the cover crop have lost their food source and are in disarray after mowing and tillage. New sets of microbes, for the new season ahead, wake up and wage war, or make peace, with the old ones, and then take over. After this rest period, the soil responds better during the next tillage, with roots letting go and decomposing materials busting up more easily.

As soon as possible the field needs to be re-sown, because bare ground is detrimental for humus production. If I don't want to plant the field with a cash crop directly, I will temporarily sow it in buckwheat to keep live roots present, shade the soil, and smother sprouting weeds. Although weeds and grass would sprout on their own and act as cover crops, it will be easier to get rid of the buckwheat.

Besides buckwheat and Sudan grass, other summer cover crops include legumes such as field peas, soybeans, and crotalaria. Deer pressure prevents growing legume cover crops on the pastures. We have to grow our vegetable crops and legumes inside an 8-foot-tall fence to keep out the deer. Don't get me started on having to feed the King's deer. They are a major obstacle for crop production throughout the Eastern hardwood forest's farming regions.

Crimson clover is a great cover crop choice in our region. We plant it between August 15 and September 30. It likes a nurse crop to help it get established. Clover is slow to get established, and when planted by itself it tends to be overtaken by weeds. Crops to "nurse" it early on include buckwheat and brassicas such as daikon, turnip, kale, or bok choy. These can also be inter-sown right before the last cultivation of a cash crop, so when the crop is harvested, the clover cover is already growing. After the end of September, we change the cover crop mixture to wheat and Austrian peas or cereal rye and purple hairy vetch, because, as I've noted, diversity in cover crops helps support a diversity of microbes. If I plan to plow and plant a field early the following spring, I don't plant a cover crop in the fall. Instead, I spread compost and simply leave the field rough plowed. Soil structure is enhanced by the pulverizing effect of rainwater in soil clods freezing and thawing during the winter, which leaves the soil more open. All other fields get tucked in under a winter cover crop.

Mulching is another way to build soil humus, although it is a bit like robbing Peter to pay Paul because the mulch usually comes from somewhere else on the farm. But we make sacrifices for the crops that supply our livelihood. Our mulch is usually leftover hay or hay that's been rained on. Sometimes neighbors offer leaves in the fall, which we use to mulch the berry patches. Leaves are better if they're chopped up, so they don't pack together in layers and create anaerobic conditions.

The group of practicing farmers who attended Steiner's agriculture course had several tasks, one of which was to determine what crop rotations build up rather than deplete humus. I had always rotated crops for reasons like pest control, or because they use different nutrients. But as I gradually came to understand soil as alive and living in time, I more and more appreciated the need for variation in plant cover to maintain humus. Crop rotations played a major role in farming historically, with priests or wise ones overseeing how rotations would be practiced to keep the land in good condition. The ordinary farmer did not make these decisions. After determining what crop rotations kept the land fertile, these rotations were rarely deviated from, as if they were written in stone.

Stricter crop rotations have really improved the soil on our farm. We follow an acre of potatoes with mixed vegetables (in their own rotation), then sweet potatoes followed by butternuts. In four years not only are the potatoes back in the first field, but that field seems to be expecting them. I wonder if soil has memory. Anyway, each time we change crops, a whole new set of microbes frees up a whole new set of nutrients.

Forests build humus, too, but it's a much slower process. Leaf mold, rotten logs and branches, and any woodland humus are a wonderful addition for gardens. Making our land available for dumping the wood chips resulting from roadside clearing has given us, ten or more years later, a nice humus addition to the compost piles. There is no such thing as organic waste—it's just more composting possibilities.

Soils with a high humus content feel silky and soft to the touch, like the way talcum powder makes your hands feel. Once you grow crops with humus, you will never want to farm without it. As Steiner stated in his fourth lecture, farmers cannot count on nature to replenish humus in soils, and this returns us to the importance of proper manuring.

> *In many districts, we cannot reckon upon Nature herself letting fall into the earth enough organic residues, and decomposing them sufficiently, to permeate the earth with the requisite degree of life. We must come to the assistance of plant-growth by manuring the earth. . . .*
>
> *We must know how to gain a kind of personal relationship to all things that concern our farming work, and above all— although it may be a hard saying—a personal relationship to the manure, especially to the task of working with the manure.*

Steiner offered us a way to gain this personal relationship with the manure by making and using some preparations.

— CHAPTER EIGHT —

Preparations

According to Steiner, the greater part of what we eat daily gives the body the living *forces* it contains, not the *substances*. Only the brain and nervous system receive substance from the food we eat. The rest of the body takes in substance from the air, through our sense organs, skin, and by breathing. This concept is hard to swallow. Is it possible that the substance of my big toe comes from what I've seen, touched, and breathed during the day, not from what I've ingested at the dinner table?

I'd heard of gravitational, magnetic, and nuclear forces, but I had always understood forces to be more or less mechanical. The idea that forces could have life was new to me. There are forces involved in life processes, such as the energy in growing plants and animals, which rocks don't have. A healthy plant or animal appears to have more energy than an unhealthy one. Maybe this is because it has more living forces.

This beginning of the lecture introducing the preparations plays a large role in Steiner's ideas on nutrition and medicine. Steiner contended that the atmosphere contains all the nutrients we need in a highly diluted state. Food grown in humus-rich soil will have living forces in it because the soil has living forces, too. This food strengthens our sense organs, Steiner said, so those organs will be more capable of receiving substances from the homeopathic amounts in the air, which they then deposit in our body.

This reference to homeopathy is echoed in Steiner's admiration of Lily Kolisko's work with smallest entities, published later as *Agriculture of Tomorrow*. Samuel Hahnemann, founder of homeopathy, believed the cure for a disease would be the very substance that produced the symptoms of that disease in a healthy person. Homeopathy uses trituration, succession, and dilution to potentize and dynamize tiny amounts of the medicine, making it exponentially more effective. In a similar way, Steiner recommended stirring to potentize and dynamize the preparations described in lecture four.

Rather than trying to find out how the production of crops can be made financially most profitable, Steiner said that the most important point was that when the crops feed us, they should not merely fill our stomachs but also further our inner life, the end view being the best sustenance of human nature that's possible. "We must vitalise the *earth* directly, and this we cannot do by merely mineral procedures. This we can only do by working with *organic* matter. . . ."

If we try to grow plants in dead soil, the food those plants produce will not have as many of these living forces. If we eat that food, our sense organs will be less capable of receiving homeopathic substances from the air, and we can suffer from nutrient deficiencies.

Farmers come to the assistance of dead soil by applying sufficiently decomposed organic residues, such as compost, to permeate the earth with the necessary degree of life. In other words, we need to pour vitality and life in all directions to make the whole farm work as a totality. This is the principle underlying biodynamic preparations—they add living forces to compost and soil, and eventually our food.

* * *

We begin with the horn manure preparation, which makes use of a cow's horn. Steiner compares a cow with a male deer. A stag is beautiful, nervous, and quick, with bone-like antlers directing currents of forces outward. Deer aren't beautiful when they're eating my crops. In contrast, a cow horn is a hollow skin/hair formation that directs forces back inward into the cow's stomach. My cows, contentedly chewing cud, are anything but nervous and quick, and could care less about what goes on around them unless they have to. Their manure is the ideal fertilizer and is what Steiner recommended for the horn manure preparation. Steiner described the process for making horn manure in his fourth lecture.

> *We take manure, such as we have available. We stuff it into the horn of a cow, and bury the horn. . . .*
> *Throughout the winter—in the season when the earth is most alive—the entire content of the horn becomes inwardly alive. . . .*

Use one hornful of this manure, diluted with about half a pailful of water. . . . Stir quickly, at the very edge of the pail, so that a crater is formed reaching very nearly to the bottom of the pail, and the entire contents are rapidly rotating. Then quickly reverse the direction, so that it now seethes round in the opposite direction.

Do this for an hour.

Sounds easy enough. The first time I tried it, I buried the horn open side up. It filled with water. Hugh Lovel was there when I unearthed the anaerobic slime inside the horn, and we had a good laugh at my first lesson. In 1992 I showed my horn manure to Alex Podolinsky. "I would use this," he commented dryly, "if it was the only horn manure left on Earth." The horns, filled with fresh manure from a lactating cow, need to be buried open side down so they don't fill up with water. They should be placed in the fertile layer of topsoil, which should not get overgrown with plants whose roots will go after the manure inside the horn. Hugh Courtney encouraged me through many years as my preparations gradually improved. "C minus," he would say and smile. The manure should turn into a humus-like substance, totally transformed from the fresh manure we make it with. As an experiment, I stuffed a pint jar with manure and buried it with the horns and it came out unchanged, still fresh and stinky.

Steiner said enthusiasm can call forth great effects. He compared the beneficial effects of a medicine a doctor specially prepares for the patient with the store-bought one that largely loses its influence without that personal relationship. Stirring the preparations with enthusiasm by hand, where even your feelings enter into it, is different from stirring with a machine. I have found enthusiasm to be the enjoyable outcome of doing what you want to be doing. We stir and apply the preparations by hand, too, although we have used mechanical stirring machines and spray rigs in the past.

As I have noted, I was quite skeptical of biodynamic preparations when I first read about them. Having been assured they did not require faith for them to work, I tried the field sprays and the Pfeiffer compost starter, not really believing anything would happen. Our first experience only became tangible the next spring when we noticed how well the potatoes had kept through the winter. Thus began the annual ritual of making and using the preparations.

I jumped in wholeheartedly and got a 50-gallon crock so I could stir 25 gallons at a time, enough for about 8 acres. After an hour of stirring, I put about 3 gallons in a bucket, then walked over an acre, dipping a whisk broom into the bucket and flinging it over the field. I spent a lot of time doing this on our pastures and gardens. Now I know that some of that time would have been better spent mob-grazing cattle instead of the set-grazing we were doing, re-mineralizing the soil with better-quality lime, and aerating the fields by ripping along contours and keyline plowing. Although we noticed some improvement due to the preparations alone, it really kicked in when combined with all the other good farming practices.

* * *

Steiner recommended following up the horn manure preparation with another one made using silica. Grind quartz to a fine mealy powder, add enough water to make a mush, then fill a cow horn with it and let it spend the summer in the earth. Dig it out in late autumn. In the following spring and summer, add a small quantity to water, stir it up for an hour like the other preparation, and sprinkle the plants externally.

I first made horn silica with Harvey Lisle in 1987. Silica doesn't undergo a visual change while it sits in the horn like the manure does, so I didn't know if I had prepared it correctly. We find geodes in our creeks, so we use the crystals from them. I have used crystals from the Smoky Mountains and the Rocky Mountains, too, along with ones from Arkansas. I hammer them into pieces that fit into a T-post tamper, and then I smash them with a big iron tamping bar. I then sift out any remaining large pieces and further grind the quartz powder between two plates of glass. You will want to wear earplugs if you do this, and be sure not to inhale the dust. Silica grinding usually happens in April or May, and the horns are dug out in November.

The horn manure preparation makes a little sense if you've experienced the remarkable fertilizing value of manure, and you can appreciate homeopathy, but the quartz preparation baffles me. After stirring a teaspoon of powder in water for one hour, the insoluble quartz still settles to the bottom of the crock. It's hard to see how it's going to help the plants. But the proof is in the pudding. The vegetables have excellent flavor and store exceptionally well.

There are also preparations that are meant to be applied to compost piles, and Steiner talked about these in the fifth lecture. "Heaven provides silicic acid, lead, mercury, and arsenic—provides them freely with the rain. On the other hand, to have the proper phosphoric acid, potash and limestone content in the earth, we must till the soil and manure it properly." These former substances are the most important of all, Steiner said, but through prolonged tillage and unwise fertilizing, we can prevent the soil from receiving them. When this happens, we should add living forces to the manure and compost, not just substances. This we will do with preparations made using common herbs and flowers.

The first preparation to add to the compost heap helps the atmospheric elements that make up protein join up with the potash salts in the earth. Take fresh yarrow flowers (*Achillea millefolium*) and clip off the florets at the end of the stems. They can be dried, to be moistened and used later, or stuffed directly in the bladder of a stag, which is then sewn shut. It is hung in the sunshine through summer and buried in good soil in autumn. After digging up the bladder in the spring, add a small amount of the rotted humus product from inside the bladder to a pile of manure or compost. This preparation gives the manure the power to quicken the earth so that it receives silica, lead, mercury, and arsenic in the finest homeopathic quantities.

Yarrow grows wild here in Tennessee, as it does in many areas. It flowers in spring. Local hunters supply us with the bladders from bucks in the fall. So we either freeze the bladders in water, or blow them up like balloons (using a straw to blow through) and tie them off and dry them. Yarrow flowers can be frozen or dried, too. I like to make this preparation in late May. I wrap cheesecloth around the bladder before hanging it outside, to keep birds off. I have made it in November, too, pouring the urine out of a fresh bladder and filling it with florets I have rehydrated with yarrow tea. With this timing the filled bladder only hangs a few weeks in the sunshine before it is time to bury it.

I wrap a piece of nylon screen around the yarrow-stuffed bladder when burying it. I mark the spot with bricks, but I have yet to meet someone who makes these preparations who has not lost the spot where one was buried. This includes me a few times, and I suppose that's part of the initiation.

* * *

We must also make the manure able to bind together calcium compounds and influences and draw them into the organic process. Steiner chose German chamomile (*Matricaria recutita*) for this purpose. We pick the flowers and stuff them into cow intestines, much like making sausages. In fall they are buried in soil as rich as possible in humus and the preparation is added to the manure after retrieving them the next spring.

A wide-toothed comb helps when harvesting chamomile flowers. Their sweet aroma completely contrasts with what we encounter in the intestines. Yarn or jute is used to tie off one end of a foot-long section, and I squeeze moistened flowers and stuff them in. I push it tight with my thumb, and you eventually learn how to not push too hard, which can cause a blowout. I fold the end over and it dries and seals up. Sometimes we hang this (and the dandelion preparation described later in this chapter) out in the summer sun, like we do with the yarrow. The appendix in the 1993 edition of the *Agriculture Course* includes the notecards Steiner made for the lectures, and there is a note about hanging the chamomile. He did not mention it in the lecture and died soon after presenting the agriculture course, so nobody was able to receive more directions. Possibly because of that, there are as many ways to make these preparations as there are folks making them. I like to build a snowman on top of where the chamomile is buried.

*   *   *

Steiner called stinging nettle (*Urtica dioica*) a jack-of-all-trades. It carries sulfur, potassium, calcium, and iron within it. I strip the leaves off the stems in late May and bury them in clay tiles with pieces of nylon screen duct-taped on both ends to keep earthworms out. We dig a pit and line it with leaf mold or peat moss, and the tiles are covered up with that before finishing with soil. I leave the tiles to sit for sixteen months, so they spend two summers underground. The finished nettle preparation is also added to the pile, giving the manure the ability to make the earth into which it is incorporated intelligent. The manure will not "suffer any undue decomposition or improper loss" of nitrogen.

We eat the nutritious nettle itself, and it transplants easily. Although a related species called wood nettle (*Laportea canadensis*) is native here in Tennessee, Steiner did not recommend substituting any other plant for making this preparation. Touching the foliage does sting, so we wear gloves when harvesting. Unescorted guests, wondering (correctly) if the nettle is a mint of some kind, can get an unpleasant surprise when putting their noses into it.

* * *

To prevent plant disease, we are advised to put white oak (*Quercus alba*) bark, ground into a meal, in a cow skull and bury it in a marshy place. Again, we leave this preparation in place over winter, and we apply it to compost piles. I use a fresh skull, within a few hours after slaughtering, or just after a cow dies on the farm. This is when we gather the intestines and mesentery, too.

I use a long narrow spoon to scoop out the white brain matter. The bark is scraped off a mature oak tree with the claw of a hammer and then ground in a Corona grain mill. I moisten the bark meal and press it tightly into the skull cavity, seal it up with a rock or bone, and bury it in a mucky place next to a pond. Oak bark ash is 77 percent calcium, and the skull is made of calcium, too. To have a healing effect, calcium must remain within the realm of life, and you'll notice that all these compost preparations are made from plants and animals.

* * *

When beginning the description of the next preparation Steiner introduced alchemy and the transmutation of elements.

> *Under the influence of hydrogen, limestone and potash are constantly being transmuted into something very like nitrogen, and at length into actual nitrogen. And the nitrogen which is formed in this way is of the greatest benefit to plant-growth. . . .*
>
> *Silicon, too, is transmuted in the living organism—transmuted into a substance of great importance, which, however, is not yet included among the chemical elements at all.*
>
> *Now in the plant there simply must arise a clear and visible interaction between the silicic acid and the potassium—not the calcium. By the whole way in which we manure the soil, we must quicken it, so that the soil itself will aid in this relationship. . . .*
>
> *We must now look for a plant which by its own relationship between potassium and silicic acid can impart to the dung—once more, if added to it in a kind of homeopathic dose—the corresponding power. . . . It is none other than the common dandelion (*Taraxacum officinale*).*

We can gain the surrounding forces by a similar treatment as in the other cases. We gather the little yellow heads of the dandelion and let them fade a bit, place them inside a bovine mesentery (the side that would be in contact with the cow's stomach or intestines), sew it shut, and lay it in the earth throughout the winter. We use the net mesentery that surrounds a cow's stomach, but I have also used the membrane surrounding the intestines. Dandelions take three days to fully open, and we gather them early on the first day, preferably before bees pollinate them. Bees and other insects flit through the air with homeopathic traces of silica following them. They touch the flower with its open ovaries awaiting pollination, and we want to beat the bees. We dry the flowers, then rehydrate them with dandelion tea when we make the pillow with them.

The first time I made a dandelion preparation, I left it in the hallway of the barn for a minute, long enough for a dog to snatch it away. I successfully buried the next one, but never found it again, despite digging a grave-sized hole looking for it. More initiation, I suppose. I finally did succeed, but my learning curve was steep. Big rocks piled on top of the spot where it's buried mark it well and prevent animals from digging it up.

This preparation gives the soil the ability to attract silicic acid from the atmosphere and beyond, which Steiner said is just what a plant needs to make it sensitive to all things and able to draw to itself all that it needs. Soon after I read Steiner's words about dandelion and was pondering how a plant could draw to itself all that it needs, I sat down in a young corn patch. There is no way this corn is going to attract nitrogen and phosphorus if I don't fertilize it, I thought to myself. Just then a bird flew over and dropped its nitrogen-and-phosphorus-rich poop, too close for comfort. It did get my attention.

* * *

Valerian (*Valeriana officinalis*) flowers are pressed into a juice with a wheatgrass juicer and then fermented for a month in a jar with an airlock attached, like what we use when making wine. A few dropperfuls of the fermented juice are added to a gallon of water and stirred for twenty minutes. We pour half of that liquid into a hole in the compost pile, and sprinkle the other half on top. This preparation stimulates the manure to behave in the right way in relation to what Steiner called "the 'phosphoric' substance." I like my manure to behave well. Steiner assures us that,

> *All this requires a certain amount of work, it is true—yet if you think it over, after all it involves less work than all the devices that are pursued in the chemical laboratories of modern agriculture. . . .*
> *With the help of these six ingredients [yarrow, chamomile, stinging nettle, white oak bark, dandelion, and valerian] you can produce an excellent manure.*

This has certainly been my experience using this method of fertility over almost four decades of market gardening. It doesn't cost much or take too much time, and our crops look healthy, produce abundantly, and taste pretty good. I've been adding these six compost preparations to the horn manure since I learned about this method in the 1990s. I'm very happy with the results. It's a way to get them all over the farm. Five-gallon crocks of preparations are stored in our root cellar. They are surrounded and insulated by peat moss, and a pillow of peat moss covers them. This is to keep their forces protected from outside influences. There is no electricity in the cellar for the same reason.

* * *

Steiner also had suggestions for problems with diseases, weeds, and pests. Moon forces affect not only the water in the ocean tides, but also the water in the soil. Horsetail (*Equisetum arvense*) is recommended for damping down too much moon force in the soil when it rains a lot. We make a tea by simmering a cupful of dried horsetail in a gallon of water for half an hour, and then we let the tea ferment in a crock for a few weeks. It smells lemony. I like to add a little of this tea to the compost piles. I dilute a quart of tea with nine quarts of water, stir it for twenty minutes, and then sprinkle it on the crops. Horsetail ash is 90 percent silica, and silica is a desiccant. This helps to explain why Steiner recommended it for preventing rot, blight, and rust. But it may not be anti-fungal as much as it is pro-beneficial-fungi. Good fungi keep bad fungi away.

Just as water increases fertility, fire can destroy it. Steiner described a method of burning seeds or rhizomes of a particular weed and sprinkling the resulting ash to rid a field of that weed. He suggested this may need to be done four years in a row to get the job done. I put seeds or rhizomes in a Dutch oven and let it sit in the woodstove all night to turn into ash. I haven't ever gotten past a few years with this technique, so I can't really say I've tried it fully, but the weeds in my field showed no decline after two years of spreading their burnt ashes. I like the theory, though, and will try it again. Steiner said that such things were known and mastered once upon a time by instinctive farming wisdom. I can see where it would be really handy to be able to rid the land of the weeds among the plants you want to grow.

"Peppers" to repel rodents or pest insects are made by burning the animal hides or the whole insect at specific times of the year. I've tried this too, and I felt that sprinkling these peppers on the farm did help a bit, but I felt uneasy burning mammal skins during an astronomical event. My experiences were inadequate to say for sure, but things could have been different had I not used the ashes. Again, I did not continue the practice for the full four years.

These ways of looking at farming problems really intrigue me. I feel confident other farmers, more dutiful than I, have had better success with them. They certainly call for more research. Steiner gave us so much to think about and experiment with. Farmers will have to pick and choose what to delve into more deeply. His interesting insights and indications leave lots of room for individual interpretations, so we are left to fathom them with complete freedom. Far from following recipes, we must figure out, on our own unique farms, how to use the guiding lines Steiner offered to find our own individual way.

— CHAPTER NINE —

Guiding Lines

"The guiding lines we shall have to give will be such that *we* can only begin on the basis of the answers we receive from you," Steiner said in the address after lecture three. He needed to know, from the farmers' description of their farms, "What is the nature of your soil, what kind of woodland there is and how much, and so on; what has been grown on the farm in the last few years . . . which, after all, every farmer must know for himself if he wants to run his farm in an intelligent way—with 'peasant wit.'" The phrase "guiding lines" pops out for me. Farmers are every bit as unique as their farms are. Consequently, during his lectures, Steiner could only indicate general guiding lines, which his listeners could then develop later for individual treatment according to the particular conditions on each farm.

At the end of the course, Steiner said he could go on offering many individual guiding lines as the foundation for many experiments. We sure wish he had, but he passed away within the year after giving the agricultural lectures. Consequently, it is up to us to follow his guiding lines as best we can, from our own perspectives, and this can lead us in many different directions. With so many different ideas, one farmer simply can't research deeply into every one. I chose certain ideas to pursue. Other farmers are inclined to choose different guiding lines.

Because biodynamics is based on experimenting with guiding lines, there are no dogmatic rules. I know farmers who make and use all the preparations, but also practice conventional agriculture. I've met some who faithfully plant by the signs but use no preparations. Some farmers measure every ingredient precisely and stir with reverent intentions, while others just grab a handful of horn manure and drink a beer while they stir. Farms I've visited in the tropics use different but similar plants when making preparations. Social aspects of farming are very important for some, and inconsequential for others. Some are card-carrying members of the Anthroposophical Society, while others have never read any of Steiner's work. I doubt I have ever met two farmers or gardeners who practice biodynamic farming the same way, and I like that.

Steiner referred to his philosophy and scientific work overall as anthroposophy. In the preface to the *Agriculture Course*, Dr. Pfeiffer notes that there was such a germinal potency in Steiner's indications that just a few sentences often created the foundation for a farmer's or scientist's whole life's work. Pfeiffer directed a composting operation in Oakland, California, that successfully recycled the city's biodegradable wastes in the early 1950s, and he left a legacy of books and lectures. Ernst Marti, Ernst Hagemann, and Guenther Waschmuth were students of Steiner's who wrote interesting books about the forces involved in plant growth, as did Jochen Bockemuhl and Gerbert Grohmann. Lily Kolisko followed Steiner's lead into homeopathy, and Maria Thun delved into astronomy. Karl König, Rudolf Hauschka, and many other early students of Steiner spent their careers following a few "guiding lines." The fascinating works of these anthroposophical scientists, and many lectures and books by Steiner, his students, and other authors, explore the deeper energetic and spiritual aspects of these eight lectures. His courses on medicine and science are particularly relevant for understanding the *Agriculture Course*.

The first time I really studied the *Agriculture Course* was when I was twenty-nine years old. Imagine my surprise when I read the following.

> *The farming anthroposophist no doubt, if he is idealistic enough, can go over entirely to the anthroposophical way of working—say, between his twenty-ninth and his thirtieth year—even with the work on his farm. But will his fields do likewise? Will the whole organisation of the farm do likewise? Will those who have to mediate between him and the consumer do likewise—and so on and so on? You cannot make them all anthroposophists at once—from your twenty-ninth to your thirtieth year.*

Saturn takes between twenty-nine to thirty years to go around the sun, so this was what they call my Saturn return. I was certainly idealistic enough.

I see now, although I dove in head over heels myself, my fields and farm didn't. I was on a steep learning curve. Besides the biodynamic aspect, I was still trying to figure out basic farming principles. There was so much more to learn, even after living on a farm my whole life, and my farm was anything but organized. I felt sorry for the grocers who wanted to know what to tell their customers about biodynamics and how the beautiful produce I was supplying had been grown. After listening to my crazy-sounding, cosmic explanation, I'm sure they wished they hadn't asked.

In 1988 I decided to become certified biodynamic/organic. Three different reasons motivated me. First, I wanted to support organic efforts in the marketplace, which were still struggling in the 1980s, and certification helped us sell produce. Second, I was alone in my early studies and wanted an experienced biodynamic farmer to visit and explain things to me about making and using the preparations.

Becoming certified meant that I had to keep detailed records of how I managed each of the fields, and get approval from the certification agency, Demeter. I dropped certification in 2002 as I no longer needed it for marketing purposes, but my biodynamic practices have only gotten stronger since. The third reason for getting certified when I started out was that very few Tennesseans had ever seen the word "biodynamics," and I was out to change that. The health food stores advertised our produce as biodynamic, and I became sort of a Southeastern cheerleader for this method of farming. I learned gradually how to explain the whole farm concept, why cows were necessary to grow vegetables, and a bit about homeopathy for the land.

I noticed the many differences in the ways that various farmers and writers presented biodynamic concepts. Eight guiding lines were my starting point in writing the first eight chapters of the book. After revisiting them again here, I'll discuss some other guiding lines from the course.

1. Rely on biological interactions to supply nitrogen and avoid fertilizers that contain nitrates or other chemicals.
2. An individual farm entity is healthiest when it produces its own feeds and fertility, and it needs livestock to do so.
3. Social gatherings on farms are important for farmers and the wider community, as is a needful dose of humor.
4. No one knows a farm like the farmer, and scientific knowledge should be augmented by peasant instincts and insights, which we can develop again ourselves.
5. Ninety-five percent of the substance of a plant is nitrogen and what Steiner calls her four sisters—carbon, oxygen, hydrogen, and sulfur—the elements that come freely from the atmosphere.
6. Silica and lime play dominant roles in agriculture, and the joining together of these and other earth elements with the air elements can be enhanced by good farming practices.

7. Building and maintaining a humus-rich soil teeming with microbes is the easiest way to raise high-quality crops and livestock.
8. Homeopathic preparations can help repair the damage done by continual farming and enrich our manure and soil to grow the best possible sustenance for the human being and human nature.

Other guiding lines are certainly worth mentioning briefly. Looking at the bigger picture, Steiner said that "everything which happens on the Earth is but a reflection of what is taking place in the Cosmos" and "that which is imaged in the single plant, is always the image of some cosmic constellation." Although these mind-expanding quotes made a lasting impression, guiding my conception of the world, I can't say they've contributed much to my farming.

As discussed in the lectures, Steiner told his listeners that the outer planets work with silica forces that ray into the Earth and stream upward with the help of clay. This affects nutrition, whereas the growth and reproductive qualities of plants are affected by the inner planets, which work with limestone forces. I picture the sun flying through space with a vortex of planets following it. Carried by sulfur, influences of hydrogen stream in from the far reaches of the universe, gathering influences of nitrogen that come from the constellations of the zodiac. These influences then join with oxygen forces in our solar system, and then join with carbon here on Earth through photosynthesis. The sun is moving in one direction through our galaxy, which is speeding along in another, while we turn with the Earth's rotation as it circles the sun. No wonder my head spins. Although I can visualize and describe these forces, I have yet to figure out how to use these guiding lines on our farm.

Although Steiner implied that planting by the moon signs can be valuable, he also pointed out that the moon "gets over our human errors." If we plant at the wrong time, forces will simply wait within the Earth until the next full moon. Nature is not so cruel as to punish us for our "slight inattention and discourtesy to the moon in sowing and reaping." Although there is ample research verifying certain astronomical consid-erations in farming, the guiding line here also includes the rationalization that allows me to relax and not get hung up with planting by the signs. Besides planting by the signs, we can help bring starry influences to our crops by cultivating in the appropriate sign. It's easier for market gardeners to time cultivation tasks by the signs than to wait around for a particularly favorable moon sign for planting. We have a little more leeway time-wise when it comes to cultivation. I always have a calendar that includes the moon and the star charts, and though I look at it frequently, I don't follow it reli-giously. I do love to gaze at the night sky and wonder.

Lunar forces affect water, as evidenced in ocean tides. Warmth makes effective the silica forces. Things really grow when it rains before a full moon, and don't grow well when it's cold outside. I like to sow before a full moon and wait until the ground is thoroughly warm before I plant summer vegetables.

According to Steiner, you need to plant oak trees in the proper Mars period, and coniferous forests in the right Saturn period. Firewood taken from trees planted without understanding these rhythms will not provide the same health-giving qualities as wood from trees planted intelligently. I get oak firewood from logging in my neighborhood, but I'm sure the trees weren't planted, let alone by someone thinking about Mars. Consequently, they probably sprouted in the right sign.

* * *

Raised-bed gardening got the thumbs-up when Steiner pointed out that it is easier to permeate the earth with humus if we erect mounds. Many backyard gardeners set up raised beds with boxed-in sides. My experience has been that raised beds dry quickly and require irrigation, which I don't have. Any piles of plant residues left on top of the garden attract earthworms, fungi, and other life enjoying the extra air, which helps decompose the raw organic matter. We always pile up the materials for our compost piles.

A guiding line that helped me understand why I need to keep the soil loose is that the air within the soil is more alive than in the atmosphere. I knew legumes fixed nitrogen, but I couldn't figure out how a soil bacterium could pull nitrogen out of the atmosphere until I realized it was from the atmosphere under the surface of the soil. As inert nitrogen from above enters the soil, it meets the microbes down below capable of assimilating it. Once oxygen and nitrogen become a part of the microbes, the air underground becomes alive. Steiner mentioned warmth also being more alive underground, but I have no idea what that means other than that warmth also promotes biological activity.

Steiner's seemingly trivial observation that a living thing always has a skin, separating the insides from the outside, got me thinking. Our 270 acres is a living entity, we are told, so the boundaries could be considered as its skin. A skin is recommended for the compost pile, but the outside of the pile can naturally dry out, forming its own skin of sorts. Even individual plant cells have a cell wall, which acts like a skin. All skins contain silica and separate life, inside the skin, from the world at large outside of it. We use the membranes of animal organs to enclose herbs for some of the preparations. The skin is the boundary for all living things. Life arises from boundaries. Think of the seashore or forest edges where so much life accumulates, or that "life always proceeds from the entire universe."

In relationship to Steiner's guiding line that enthusiasm produces great effects, he noted that it makes a difference whether farmers shake the seed corn in their hands a bit as they're sowing. I liked this observation and continued to hand-sow 6 to 10 acres of vegetables for decades, refusing to buy a planting machine. We enjoy walking the furrows, dropping and stepping on the seed, at least for the first few hours. I don't necessarily shake the seed but am certainly more conscious of my actions than if I were on the tractor. Although not opposed to mechanical seeders, I just never did get one.

Steiner realized that machines were necessary in farming, but felt that farmers didn't need to be crazy about machines. His advice was to fix up your old machines rather than rushing out to buy brand-new ones. This fit my budget, and I liked old stuff, even though some of it could have been replaced long ago. I became infatuated with Farmall 140 cultivating tractors, which are perfect for market gardening. But they are fifty to sixty years old, so I bought some extra tractors at local auctions as a source of parts. I continue to use old farm equipment, much of it originally made for horses to pull. My goal was to make a living by farming like my neighbors were doing, with the same equipment they used, so that the only difference would be that I was doing it organically. Sometimes I wish I had more, newer, and better equipment, but it's hard to justify the expense.

*　*　*

Steiner's first guiding line for feeding animals was that they should be outside, giving them the opportunity to come into relationship with the surrounding world by sense perception in the finding and taking of their food. Then there is the guiding line that the head needs substances from the earth and the body needs substances from the air, which it receives in homeopathic doses. So he recommended feeding roots and hay to growing stock, such as carrots for the head and grass to assist it in passing through the body. For milk production we need to stimulate the middle of the animal, and a plant strong in foliage like clover is best. To fatten an animal we use seeds, fruits, and feed with what Steiner called the "fruiting process" enhanced, done by cooking, steaming, or drying (this is explained in more detail in the section on lecture eight, page 170). Even cultivation, which makes things like turnips and beets grow bigger than in the wild, he called a fruiting process. Quality salt is important, too, and we make sea salt available for all our farm animals. It contains many trace elements, which afterward are spread over the pastures in a more readily available form after passing through the cattle.

Our cows eat grass, clover, and whatever else they can find in the pastures. The more we rotate pastures, the better the cattle and the pastures look, but the more they complain if they don't get moved to better forage. I can barely grow enough carrots for the CSA, let alone for stock feed. Twice we have taken the effort to cook food for the pig, and it made an unbelievable difference. Cooking makes food more digestible. We just boiled discarded sweet potatoes, Irish potatoes, and butternuts until soft, as these are what we have a lot of. We process hogs with the neighbors, and everyone was duly impressed when we split ours open. Instead of a uniform, light pink, the insides were bursting with colors. They hadn't seen that since their families raised pigs back in the old days, when preparing hog slop was a daily chore. Salted, smoked, and hung for five years, it made for the best of meats.

* * *

Many in my generation felt the need to do something to save the world from various issues we thought were detrimental to society, the environment, or the future in general. These feelings and thoughts led me to organic farming and renewable energy, while others followed different paths. It's easy to complain, harder to do something about it. I don't know what conditions were like in Steiner's time, but this comment from the discussion after lecture six sounds similar to today.

> *Admittedly, when we consider certain phenomena of our time, we might become a little pessimistic; but in regard to this question of the moral improvement of life we should never tend to a mere contemplation of facts. We should always try to have thoughts that are permeated with impulses of will. We should consider what we can really do for the moral betterment of human life in general.*

In another context, Steiner noted that every idea that does not become your ideal slays a force in your soul, and that every idea that becomes your ideal creates life forces within you. Nowadays, we say this as "walk your talk." Eating from our own farm seems to help me. Steiner mentioned poor-quality foods, resulting from unwise farming practices, as the reason why folks who held genuine, heartfelt feelings and thoughtful ideas might not be able to implement them in practical life. He called it a lack of willpower that could be corrected with better agricultural methods. Unfortunately, these have not improved since 1924.

Many of Steiner's guiding lines seem to be about raising human consciousness, along with principles for better farming. In lecture seven, he remarked on deepening the experience of the sense of smell, saying, "You can easily become clair-sentient with respect to the sense of smell, especially if you acquire a certain sensitiveness to the diverse aromas that proceed from plants growing on the soil, and on the other hand from fruit-tree plantations—even if only in the blossoming stage—and from the woods and forests!" He also advised his listeners to distinguish, to differentiate, and to individualize between the scents of herbaceous plants and the scent of trees. The air smells thinner in a forest, like it does at higher altitudes. The meadow smells more down to earth, a bit more comforting. I like this practice because I love taking walks in the woods. Our farm consists of meadows and pastures interspersed throughout a large, mature hardwood forest. I'm glad I was encouraged to develop a sense for the different aromas I encounter daily.

Clairsentience is the supposed ability to discern what we don't perceive under normal circumstances. I believe the nose is one of the best scientific instruments for learning about what's going on around the farm. For example, the smell of a sick animal's manure is more unpleasant than that of a healthy one. I tell folks to not put something in their garden if it smells bad. The ammonia smell means nitrogen is escaping, and we don't want those nitrates. It is much better to fully decompose everything before applying it to the garden.

Living off-grid for twenty-five years affected me in more ways than I know. Without electric lights, I felt closer to nature, following the daily and yearly rhythm of light and dark. Perhaps it's because I was younger then, too, but everything was simpler and less confusing. Although Steiner grew up without electric power in the home, I imagine he lived around it as soon as it was available. He didn't have good things to say about it during the last discussion, though. "Human beings cannot go on developing in the same way in an atmosphere permeated on all sides by electric currents and radiations. . . . Steam works more consciously, whereas electricity has an appallingly unconscious influence; people simply do not know where certain things are coming from."

Influenced by others in the counterculture, Debby and I, along with many others, became vegetarians in the mid-1970s, for health, environmental, and economic reasons. But we still kept the cows because of the value of their manure and to maintain the pastures. Steiner's insistence on cattle for farms might seem ironic in light of his statements about vegetarianism. He thought that refraining from meat makes us stronger, because we draw forth from out of ourselves that which would otherwise be lying fallow or dormant. It all depends on the individual, because some will benefit and it's not right for others. After a decade or so, I began eating meat again because it was recommended for people with my blood type. I didn't feel any different after making the change, but it really cut down on the quantity of carbohydrates I was eating.

Permaculture thinking abounds in the alternative agriculture movement, and we find Steiner pioneering these concepts in 1924, long before the word was even coined. He observed that nature is wiser than humans, and that a forest has value for the surrounding farmlands, particularly where the land is naturally wooded. Birds and mammals love the trees and shrubs, and the regulation of forests is an essential, significant part of agriculture. So I let hedgerows and cedar trees grow up on slopes that had been previously cleared. That certainly benefited the wildlife, especially for some of my worst garden pests, such as deer, groundhogs, raccoons, and crows. But it is well worth having woods surrounding the gardens because the biology generated in the humus-rich forest floor finds its way throughout the rest of the farm by all that wildlife traffic. Just in the last few years we have witnessed the return of beavers, river otters, mountain lions, and bald eagles, and black bears and wild boars are already in neighboring counties.

Meadows rich in mushrooms are also important for the farm as a whole. One reason is because of the kinship between mushrooms and harmful bacteria and fungi. The meadows with their beneficial fungi, the mushrooms, keep the harmful, microscopic creatures away from the rest of the farm. Half of our farm is forested, and I let more grow in some places and thin it out in others, trying to find a proper balance between the woods, the shrubs, orchards, gardens, pasture, and meadows. Steiner called this "the essence of good farming."

"Go on manuring as before" is one of the guiding lines I plunged headlong into. Sprinkling homeopathic doses of preparations on our eroded, worn-out, hillside soils was simply not getting enough results. Dad had a few old farming books, and they enlightened me to the holistic approach common in the late nineteenth century. I began gathering anything written on farming before World War I. Biodynamics worked way better for me when my guiding lines included lessons from hundred-year-old grade-school textbooks on agriculture.

CHAPTER TEN

Go on Manuring as Before

Steiner remarked that there was no need for him to tell farmers to "go on manuring as before," and in his time, that was probably true. The principles of farming were common knowledge. Steiner's agricultural course was not to teach people how to farm, as his listeners already knew how to do that. He was offering them a broader perspective on agriculture as a whole, as well as new insights on how to enhance the good farming methods already in practice. Once this dawned on me, I realized that to truly understand Steiner's agriculture course, I needed to learn how folks had farmed at that time in history, particularly around where I live. So began my study of nineteenth-century agriculture, which had its roots in European farming traditions.

I had help in this from my father, who was born in 1906. Like most youth in the early twentieth century, my dad left his family's farm as soon as he could, but he always retained an interest in farming. I began learning about old-time methods by reading some of his old *Yearbooks of the Department of Agriculture*, and I was struck by the depth of knowledge apparently common before his time. Our 1903 *Encyclopaedia Britannica* has a fascinating, 125-page article on the history of innovations and improvements in farming, culminating in a system that was sustainable, regenerative, and organic.

Now that my interest was piqued, I began collecting and reading anything I could find from late-nineteenth- and early-twentieth-century literature on farming. That new knowledge, similar to the wealth of wisdom my elder neighbors provided, gradually convinced me that people better understood how to raise plants and animals a century ago than they do now. How could that be? I was under the impression human knowledge was progressing. But words can distort truth as well as reveal it, and the agricultural advice from the land-grant colleges for the past hundred years runs counter to nature's laws, good farming principles, and common sense.

Nowhere is the dichotomy of agricultural education before and after World War I more apparent than in the evolution of the *Yearbooks of the Department of Agriculture*. Older ones are chock full of fascinating articles and research, plus statistics reporting the year-by-year production and acreage of various crops. From the 1897 yearbook I learned how much of the corn crop was retained and consumed in the counties where it was grown. In Tennessee, on the average, 86 percent was used in the county where it was grown, and only 14 percent was shipped elsewhere. In Kentucky, 89 percent stayed. Growing corn can deplete the soil, but this can be remedied by composting the leftover fodder with manure and returning it to the fields. At that time the USDA was promoting sustainable farming practices like those Steiner spoke of when describing the self-sufficient farm. The more modern *Yearbooks of the Department of Agriculture* paled in comparison, recommending the separation of animal husbandry and crop production and the use of products whose manufacturers fund university research.

I fell in love with grade-school textbooks from that era. In the preface to *Agriculture for Southern Schools*, John Frederick Duggar wrote that his aim was "to arouse the interest of the pupil in nature, and especially the common plants of the farm, orchard and garden, and for the student to master the subject by stimulated observation and quickened thought rather than by mere memorizing." The teaching of observational skills was fundamental in these old textbooks.

For example, how do you tell a good wheat seed from a bad one? You use a magnifying glass and look at it closely. You were required to observe what your parents and neighbors were doing on their farms and report back on why and how they did what they did. Lessons reflected what was going on in the students' lives and communities.

I remembered our class doing some of the experiments described in the old textbooks. To understand the importance of drainage, take two tin cans, poke small holes in the bottom of one, fill both with soil, plant seeds in the soil, and water. By setting up this test and observing the results, rather than just reading about the importance of drainage, the realization that seeds can't sprout in soil without air comes from your own experience.

Wrapping dark paper around a glass jar and growing a plant in it reveals the interesting root hairs when the paper is removed. After legumes are dug up to expose the nodules, they are described as tiny fertilizer factories operated by microbes. Osmosis, the process by which liquid is taken in through the thin membranes of the root hairs, is demonstrated by the following experiment. In the small end of an egg, a hole is made a little bigger than a pinhead. Over this hole a short piece of glass tubing is fastened with melted wax. A bit of shell is chipped away from the large end of the egg, without breaking the inner membrane. The egg is then placed, large end down, over a jar filled with water. The student must explain why the water rises through the thick fluid in the egg and into the tubing, even though the membrane shows no pores when observed through a microscope.

In the late 1800s, seventh graders were expected to carry a pocketknife to school, where they learned how to graft and bud fruit trees. They dissected flowers to learn how to identify the male and female reproductive organs, how to tell whether a strawberry flower was self-fruitful, and to understand the principles of cross-pollination and plant breeding.

The books informed me that plant food comes from two sources, the atmosphere and the soil. Stars and gases from outer space eventually came together to form our solar system and planet Earth. Over time, rocks decay into soil and soil hardens into rocks. Describing the slow process of lichens and mosses breaking down the rocks with the help of microbes explains how soils are formed. Higher plant forms slowly evolved, as did the microscopic animals, sometimes referred to as animalcules.

The books impressed me with photos of the beautifully plowed soils, the compost piles and lush cover crops, and the weed-free gardens and fields. The pigs, sheep, and cattle did not look real. Their huge bodies with small heads on such short legs gave them a comical appearance. One pair of pictures summed up the necessity of keeping livestock. The first showed a load of hay leaving the farm with the caption "hauling much fertility away." In the second photo, a load of dairy products was leaving the farm with the caption "hauling little fertility away." A photo of a town with good roads, schools, and churches indicated dairy country, because out of all agricultural products, milk results in the greatest long-term wealth by conserving soil fertility. Since you couldn't buy it, fertilizer was a result of farming and not an off-farm input.

The drawings were helpful, too. One picture compared soil layers being sheared by the plow blade to the way pages shift upon each other in a paperback book when the book is curled over but kept closed. This action brings air into the soil where the roots can access it. Loose, aerated soil allows more rain to soak in rather than running off the surface, and root hairs can stretch through it more easily. A drawing of soil particles coated with a film of water down to the subsoil showed capillary action. As plants use up the moisture in the topsoil, the water from the water table below rises through the film water adhering to the particles all the way up to the drier soil. A footprint on dry soil stays moist, portraying water leaving the soil into the air, while the soil not stepped on remains loose and the dry soil acts as a mulch, which conserves moisture. This is why we harrow, hoe, and cultivate shallowly after a rain, to check evaporation. The only water leaving the ground should be through the transpiration of the plant leaves. Students learned about transpiration by noticing the moisture condensing on a jar placed over a growing plant.

All farms had to have livestock to maintain fertility. I learned that English land leases required the farmer to sow the land back into a grass/clover sod for two years after cropping it for two years. You could not sell hay, only the animals raised on it. Manure was highly prized and cared for. Young men would borrow a load of manure from a neighbor before bringing their prospective fiancée's parents over, because that was a sign of wealth. Bedding absorbed the urine, and it was all fermented in compost heaps or applied directly to cover crops in autumn. Market gardeners applied 75 to 100 tons of compost per acre annually and reaped two or three crops every year.

The books said that gardens must be rich and well drained, supplied with plenty of humus, and manured heavily. The soil should be plowed deeply, and the most thorough tillage of crops practiced. A garden should never be left bare to grow up in weeds, but planted in cover crops instead, preferably containing legumes. Over and over again, the old-time books advised readers to care for the soil microorganisms, the farmer's best friends.

I learned that farmers formed associations that helped keep records, conducted experiments, and fostered clubs for the students. Grade-schoolers bred their own corn, and one boy raised 220 bushels to the acre when the national average was more like 26. Corn production per acre has steadily increased since then, and a 2011 Farm Bureau magazine proudly announced it was the biggest corn yield ever. It wasn't until later in the article that the writer revealed that it was the biggest yield *since 1917*. Of course, there was much more acreage in corn back then and a lot more people working it. But agricultural progress has been questionable and certainly problematic since that horse-drawn, manure-powered era during which they raised more corn than we do now.

The chapters on the importance of social life intrigued me. Farmers can't pick and choose neighbors and associates like town dwellers can, the books said, because of the greater distance between them. So they are much more dependent on each other, requiring mutual helpfulness, tolerance for differences, and more concern for their communities. Much of the daily work is for the long-term health of the farm and community rather than for short-term income. Farmers don't work nine to five with weekends off and a weekly paycheck, but learn what they can live without or make for themselves. The greater freedom farmers enjoy comes with greater responsibilities.

Intimate interactions with nature and observing her laws teach the farmer to become humble, honest, patient, and reliable. Farming is one disaster after another, I read, and it's a special gift of the farmer to maintain good humor when the cows get out, drought or torrential rains threaten a crop, prices fall, and machines break. I learned that happiness, like success, depends more on habits and thoughts than circumstances.

The saying "don't believe everything you read" also applies to these old farming books. Paris green, an insecticide, sounds pretty until you learn it is lead arsenate. The books advise to till often, but at least the tools they used didn't compact the soil like huge tractors and rototillers do. The books also reminded me that 150 years ago women and minorities had far fewer rights. But all in all, the discerning reader seeking historical agricultural knowledge will find a wealth of it here.

Besides the solid, practical advice, the way ideas are presented in these books can be truly illuminating. I'll close this chapter with a few quotes from *The Principles of Agriculture* by Liberty Hyde Bailey, written in 1906. The purpose of agricultural education "is to improve the farmer, not the farm. If the person is aroused, the farm is likely to be awakened. The happy farmer is more successful than the rich one." He also writes, "A book like this should be used only by persons who know how to observe. The starting point in the teaching of agriculture is nature-study—the training of the power to actually see things and then to draw the proper conclusions from them." I think Steiner would agree.

— CHAPTER ELEVEN —

Observation

What you look for is what you find. Like most of us, Steiner tried to make sense out of his impressions of nature. In *Philosophy of Freedom*, he proposed that the observer affects the observation, a hypothesis proven decades later in physics with Heisenberg's uncertainty principle. Subatomic particles look like particles if that's what one is looking for, but also look like waves if that's what one is looking for. Our concepts limit the choices of our observations, hence our understandings of what we encounter. The ease with which critical thinking can lapse into reasoning within traditional ruts of thought obscures reality, as if we are trapped by the walls of our past thoughts. Steiner set about to expand his concepts.

Editing Goethe's scientific works brought to light the importance of developing his observational skills. Just as I am a student of Steiner's work, who lived one hundred years ago, he was a student of Goethe, who lived one hundred years before that. Goethe accepted the dominant viewpoint of Newtonian science as far as the study of inanimate objects went, but found it was insufficient for the study of life. In Goethe's time, scientists described phenomena solely according to substance, appearance, and external characteristics, void of any underlying principle. Scientists studied plants and animals only after they were dead, using the same approach with which they studied inorganic things. But nothing alive and growing is static. Life is always in continuous, fluctuating movement, although perhaps changing so slowly we don't notice it. Taking a plant or animal out of its place in nature and studying it in a laboratory can never give us the total picture. This reductionist way of looking at nature runs rampant in our current age, too. Instead of taking things apart to study them, Goethe wanted to know how plants and animals fit into the whole picture.

By forming a hypothesis and summarizing the results of repeatable experiments into concepts, the scientific method has exponentially expanded human understanding of the material world. These concepts, representations of things that don't change, have led to great discoveries over the last few centuries. Live things are different, as they grow and change all the time. We can better understand animate objects by patiently and carefully observing their growth process throughout their whole lifetime. If we consciously refrain from jumping to conclusions, our repeated objective observations gradually illuminate the principles that animate life, the intelligent design of something in the process of becoming. Goethean science, as this method became known, takes time.

Luckily for me, Goethe wrote simply, and his concepts are easy to grasp, whereas Steiner seemed to try to make everything complicated and hard to understand. Goethe's *Metamorphosis of Plants* clearly describes an annual plant through its life cycle, as the young plant's crude sap becomes ever finer. By continuous study of plant growth and the various modifications a single plant species can develop in different soils and climates, Goethe found the life principle in the formative forces that direct plant growth.

For example, some species of mountain plants develop smaller, narrower leaves and more widely spaced buds when growing at high altitudes than do species growing in the lowlands. Goethe knew the principle that makes an entity a plant could not be found strictly in its appearance. The endless manifestations in the plant world have the same basic form, the primal image of an archetypal plant. The nature of a plant is the totality of mutually interpenetrating, formative forces that shape it into what we see. These forces are Earth, with its elements and life in the soil; Water, with its ability to dissolve minerals; Air, containing light; and Fire, which we experience as warmth. This division of world phenomena into these four elements is as old as the hills and found in many ancient cultures. See the table "Juxtaposition of the Elements" on page 124.

In the Plant Elements row of the table, hydrogen could be listed in the Water column, too, and sulfur in the Fire column. Oxygen is used in combinations with each of the other four elements, so maybe it doesn't belong only in the Water column. One of the things I like about Steiner and other scientists is that they refrain from either/or judgments and continually

Juxtaposition of the Elements

Classic Greek elements	Earth
States of matter	solid
Effects of elements	soil
Kingdoms of nature	mineral
Effects of kingdoms	physicality
Formative forces	life
Parts of plants	root
Elemental beings	gnomes
Plant elements	carbon
Zodiac signs	Taurus, Virgo, Capricorn

question their own answers by keeping their minds open to other possibilities.

An example of how we use Steiner's ideas on our farm is the way we treat crops and animals compared to modern methods. Rather than regarding plants as just a conglomeration of minerals grown using artificial fertilizer, we immerse them in a life-filled, humus-rich soil. Our livestock are not treated as a crop we simply buy feed for and raise in confinement, but are allowed to express their essential nature and use their senses to forage for their food outside. We might plant by the signs, following the ancient knowledge that Fire signs

Water	Air	Fire
liquid	gas	heat
moisture	atmosphere	sunshine
plant	animal	human
growth	sensitivity	consciousness of self
sound	light	warmth
leaf	flower	fruit
undines	sylphs	salamanders
oxygen	nitrogen	hydrogen
Cancer, Scorpio, Pisces	Gemini, Libra, Aquarius	Aries, Leo, Sagittarius

are a time for planting fruits, Air signs for flowers, Water signs for leaves, and Earth signs for root crops.

To understand what makes a plant or an animal grow, I needed to sharpen my powers of observation and concentration. Three Steiner books—*Knowledge of Higher Worlds*, *Theosophy*, and *Occult Science*—offered exercises for this. There I encountered virtues, conditions, and requirements such as devotion, generosity, concentration, equanimity, open-mindedness, positivity, compassion, loving kindness, and perseverance. I had my work cut out for me. I guess it made sense that my education was dependent on improving my personal character.

Understanding that the sequences of my own thought patterns were similar to the metamorphosis of plants, I observed the transformations of seeds, shoots, and leaves into buds, flowers, and fruits, and the eventual withering and dying of plants on the farm. When I concentrated on one particular observation, I followed this metamorphosis forward and backward in my mind. When carefully observing a seed, I can imagine the shoots, leaves, and so on in the seed's potential future. The difference between observing a flowering plant and a dying plant is like the stark contrast between day and night. A continuum of the ever-changing forms of a plant can be observed during its life cycle. Thinking about the plant's crude sap in the cotyledons becoming ever finer as the plant grows and eventually blooms correlates with my own crudeness as a teenager maturing into finer qualities as I continued to grow. Well, sometimes. The growth I witnessed around me in the annual cycles of crops and animals growing and reproducing was mirrored in my own thought processes. Most seeds end up being eaten or rotting, just as many of my thoughts are mere imaginations of my mind. But some inspired thoughts can lead to an intuitive comprehension of life processes, just as some lucky seeds grow and reproduce. Patience is the necessary virtue here. None of this happens quickly.

Thinking about thinking, with its continual transformation of one thought to the next, is like watching plants and animals grow and develop on the farm. Induction led me to using analogy to extend what I observed day-to-day to become aware of general

principles. This suited me better than a deductive process of assuming a general truth and seeking to connect
it with what I observed. Just as the image of a building
exists before the building itself does, there seemed to be
an intelligence in nature guiding the forms I observed.

Along with my collection of old farming textbooks,
I have old math and psychology textbooks that my dad
collected. In those books, I noticed quite a difference
in the way ideas were presented in the nineteenth century versus how they are now—they were much more
practically oriented and dependent on the student's
personal observations. These writings are remnants
of a former consciousness that depended on direct
observations of nature.

The instinctual peasant wisdom Steiner discussed
was developed over a long period of time, when there
weren't so many distractions in daily life. We can learn
to read the book of nature by paying attention to her
metamorphosis and following the growth processes
and life cycles of the plants and animals around us.
This can take years. It does not require abandoning
scientific achievements but adds a depth to them that
gives us a broader understanding. Eventually we learn
to trust ourselves as capable co-workers with Nature,
following her constant changes to finally grasp her
immutable laws.

Goethe and Steiner were not farmers, but they laid
the groundwork for an understanding of nature that
respects her laws. Their ideas have accompanied me
for half a century as I discovered the reality of stewardship, which has been way more work than I ever

imagined. Besides making a living by growing and selling vegetables and cattle, I try to create a park-like atmosphere on the farm, humanizing nature with trails to swimming holes, limestone bluffs, waterfalls, and the pristine creek flowing through fields and forests. The events, the CSA, our cows, and an apiary have made this a community farm, and folks come here to learn just as I learned from the older farming community when I came here a long time ago. I see the world through the lens of a farmer. Even the Bible seems to me to be about farming and creating a land flowing with cattle and bees—I mean milk and honey. Only because I am a farmer did I feel I could tackle the job of interpreting Steiner's agriculture lectures in everyday language, and I'm honored to have enlisted the help of fellow farmer and friend, the late Hugh Lovel.

CHAPTER TWELVE

Agriculture, Simplified

Reading Rudolf Steiner lectures is not easy, so I set out to simplify them. In this chapter, as I go through each lecture and put the concepts into plain language as best I can, I shift between a direct description of what I think he's trying to say and brief reflections of my own. Hugh Lovel collaborated with me and was helping to edit this material when he passed away in 2020. My aim is to inspire you to read the *Agriculture Course* yourself and put its wisdom into practice.

Lecture One: June 7, 1924

After gratefully appreciating the hosts of the agriculture course, the first thing Steiner points out is that those who talk about agriculture should have a sound basis in it, and really know what it means to grow beets, potatoes, or corn. He includes the social aspects, the organizational nature, and the economic realities. In several passages, Steiner describes himself as a peasant, and honors the wit, observational skills, and instincts of country people. He would far rather listen to a chance conversation with a farmer than a scientist.

Steiner's comments about the social sides of agriculture echo Tolstoy's observation that the people doing the work are of the utmost importance. Social life in rural areas develops out of families. Everyone knows each other and their peculiar talents, habits, and personalities. This allows for an equitable distribution of work and goods because it is all on such a small, intimate scale. Agriculture and civilization grew up together and remain inseparable. History suggests that when peasants move to cities, they bring with them immense practical intelligence. But removed from agriculture, education tends to lose its sound basis. How rural society takes care of itself is best left to those who live and work there. Humbleness, compassion, and practical sense become ingrained in anyone who cares for land, plants, and animals.

The same is true of the organizational nature of a farm. It seems obvious that those who are in constant touch with the land are the ones who know best what to do. Organizing for short-term profit rather than long-term abundance creates problems on farms. A comprehensive view is required to understand the many interactions that make a farm thrive. The farmer is the only one who is in the position to organize this complexity.

The economic realities in farming also need to remain in the hands of the growers. They are the ones who know how much it costs to grow the animals or crops on a yearly basis. When outsiders interfere too much, it's the farmers who end up paying the price. Supply and demand can create price fluctuations that don't reflect the costs of production, which, first and foremost, must be covered. Once the farmer is fully compensated, then and only then should others concern themselves with the price of farm products. Steiner insists that it must work in a business-like, profit-making way, or it won't come off.

Steiner's affinity with Goethe surfaces when he mentions influences coming from the entire universe affecting what people erroneously consider to be self-contained entities. A prerequisite to understanding the *Agriculture Course* is the realization that all things in nature are interconnected. A plant cannot be understood in isolation, removed from its interrelationships in the field. It would be like trying to understand why a compass needle points north without taking into consideration the Earth's magnetic field.

Farmers must study the interworking of nature. Centuries of observation have led to crop and animal rotations, the development of our crop varieties, and the secrets of building fertility. Farmers' instincts were quite specific and reliable when they were closely connected with nature.

Next, Steiner indicates what is most important in agriculture. Instead of talking about the chemical and physical components, he asks us to look carefully at how human beings live. Humans have a considerable degree of independence from the outer world, but this is less true for animals. Plants are embedded in, and quite dependent on, what is occurring in their earthly surroundings, which is a reflection of what is occurring in the universe. Plants are directly affected by the seasons, animals less so, and humans appear quite free in regard to these cycles. As we view the sun, moon, and planets over the course of a year, we see them move across the twelve constellations of the zodiac. In the old "instinctive science," the seven lights in the sky that moved differently than all the other stars had this sequence: Moon, Mercury, Venus, Sun, Mars, Jupiter, Saturn. Steiner has much to say about this planetary life and how it's connected with the earthly world.

He begins with the role of silica, the combination of silicon and oxygen that makes up one half of the Earth's crust. Considered unimportant in agriculture because of its insolubility, silica is actually of the greatest imaginable value. It is present in very fine dilution in the atmosphere, and is necessary for receiving what comes from the distant planets: Mars, Jupiter, and Saturn. On the other hand, the influences of limestone and kindred substances are affected by the inner planets: Mercury, Venus, and the Moon. Plants only thrive in the balance and working together of these two forces, in the cooperation of limestone and silica.

Although Steiner didn't have the knowledge gained over the last hundred years, he understood the dynamics of photosynthesis and plant root relationships with microbes. The way he presents this is through the great polarities in Nature, lime and silica, with the plants in between. The clay/humus complex in the soil mediates the forces of these two poles. Silica conducts the nutritive influences of Mars, Jupiter, and Saturn. Conversely, lime makes plants receptive to the reproductive influences of the Moon, Venus, and Mercury.

Warmth promotes the forces of silica. Plants need warm weather to ripen their nourishing fruits and seeds. A lack of silica will mean less nutrition. Silica is everywhere in the environment in minute doses. When plants become nourishing food or fodder, substances like silica are involved. Both the silica-coated fungal hyphae in the soil and the silica-rich tubes of the fibrovascular bundles in the plant enhance the plant's ability to efficiently access a wide array of nutrients throughout the soil.

Water promotes the forces of lime, and lime brings nitrogen into the plant. Water is the ideal substance for the distribution of lunar forces. Plant growth shoots up after a rain and a full moon. A lack of water or calcium limits the capacity for growth and reproduction.

The concentric circles of the orbits of the planets are not all in one plane, but instead form a spiral. Because the sun travels through space, there's a vortex of planetary orbits following behind it. Steiner postulates that forces from the universe enter this vortex and are absorbed by the siliceous rocks in the Earth, and then ray upward. Other forces come from the universe by way of the inner planets and are received by the Earth as lime forces raying downward.

Steiner notes that people go through life quite thoughtlessly, glad not to have to think about such things. They conceive of nature in a materialistic way, functioning like a machine. Steiner acknowledges the value of human achievements with machines, and the need for instinctual peasant wisdom to step aside for the rise of scientific discovery. He feels it is time to rejoin the two worldviews. Life does not work in a mechanical way. A living organism is not a simple reductionist system, but a very interdependent and complex set of interactions. It reaches out from microbes to stars. This is the primary lesson in the first lecture. We have come to a starting point with the revelation of how silica and lime work to bring nitrogen into our crops in the proper way for maximum animal and human nutrition.

Lecture 2: June 10, 1924

Steiner begins the second lecture with an overview of the whole agriculture course. We will spend the first lectures gathering knowledge to recognize the conditions on which the prosperity of agriculture depends, he explains, by observing how all agricultural products arise and how agriculture lives in the totality of the universe. Then we will draw practical conclusions that can only be realized in their immediate applications and are only significant when put into practice. But to start, we must gather, recognize, and observe.

Here we see Steiner's holistic and observational approach to science, compared with Newton's reductionism. We are not starting with a problem and hypothesis, as in Newtonian science. Instead, we are looking for information, conditions, and how something (agriculture) lives. There are no boundaries to where we will look.

The first condition is clear. A farm is true to its essential nature if it is conceived as a kind of individual entity in itself—a self-contained living entity. You should try to have whatever you need for agricultural production within the farm itself, including the right amount of livestock. Steiner insists that a farm needs livestock, and then explains why. Our livestock should get their feed from our own farm, and our farm should get its fertilizer from them. He states it is not the same to get manure from neighboring farms. What makes our own farm's manure different?

The humus content of soil, with its carbon, bacteria, fungi, and protozoa, is formed in large part by the animals on it. They eat what grows on the farm, and then digest, transform, and integrate the farm's microbiology inside their stomachs. A cycle of rejuvenation happens as fertility is returned to the soil and new plants grow there. Steiner justifies this need for a farm's own livestock by considering not only the Earth but also influences from the distant universe. This digestion by the livestock balances the digestive lime forces of the manure in the earth with the silica forces of information from afar. This will be considered from various standpoints, beginning with the soil.

The soil is alive and is more than the sum of its parts. We can observe that this invisible life of the earth and its "fine and intimate" activities is different in summer than in winter. We can also observe that the surface of the Earth is a kind of organ somewhat like the human diaphragm. All the plants, animals, and people on Earth's surface are living inside the belly of the farm organism. The head and nervous system of the farm are underground, in the roots and soil life, while its reproductive organs are up above. In this way, the farm is like an upside-down person. There is a continual, dynamic interaction between what is above the Earth's surface and what is below.

We are asked to observe what these interactions are and where we find them. In lecture one we were introduced to lime and silica and their relationship to the inner and outer planets, respectively. The life processes and reproductive activities associated with lime occur above the Earth and depend on the inner planets and the sun. The informative and nutritive activities of silica that occur beneath the Earth's surface depend on the outer planets and the sun.

Notice that both sets of planets work with the sun. It is through the silica in sand, rocks, and stone that we have influences from the distant universe. This is where life comes into the soil, through the communication of information by silica.

You may wonder how what is absorbed by the soil gets carried back up into the plant. The greater surface area and charge of clay particles facilitates this absorption and transportation. Adding clay to a sandy soil and adding sand to a clayey soil are old-time farming remedies because the soil needs both. Clay is the carrier of the upward stream of silica's activity beneath the soil.

Plant growth through photosynthesis in the air above the soil is a kind of outward digestion, not unlike the inward digestion of animals. A mutual interaction arises when what happens above ground is drawn downward by the limestone in the soil. Farmers spread lime on top of their fields, knowing it will work its way downward.

Steiner tells us there are two kinds of warmth for plants—a leaf and flower warmth that is dead, and a root

warmth that is living. The moment warmth is drawn into
the soil by the limestone, it is enlivened. Air, too, is alive
below the soil surface and dead above. The soil is full of
aerobic, live beings, much more so than the air. Water,
on the other hand, is more dead below the ground than
above it. By losing life it becomes receptive to distant
forces, especially in midwinter when the minerals in the
earth come under these influences. These are the crystal-
forming forces. Experiments have shown that crystals
precipitate out of solution more easily underground
in winter than in summer. Before and after midwinter,
minerals ray out forces that are particularly important
for plant growth.

To till the soil, we must know the conditions that
enable distant forces access to it. We can learn this from
the seed-forming process. When the plant matures and
its protein is the most complex, the lime and silica
forces within the plant separate, and the plant is driven
into chaos. Seeds, as they form, become receptive to
the influences of their surroundings. We prepare the
soil for planting seeds by creating the condition for a
new complexity. This is done by growing cover crops,
applying compost and minerals, and gentle tillage.
Steiner says that a plant is always the image of one of
the starry constellations, which is carried forward in
the seed. This new organism is formed from the life
forces of the entire universe, and the best way to grow
the new plant is to place it in humus-rich soil.

In most plants the silica nature is in the root and
the lime nature is in the canopy. Normally, the distant
silica forces work upward from within the earth in a

single, central root up the stalk to the flower. But in a highly divided root, as with grasses, the earthly nature is working downward from its normal place above the soil. These plants form a thatch with their extensive root system and are the fodder plants that build good soil. The best soils are found in the steppes, plains, and savannas where grasses have been grazed periodically for centuries. We mimic this by rotationally grazing grassy pastures and sowing grain cover crops. These are all silica-rich plants with sharp, pointy leaves. If we want the silica forces to remain below, we put the plant in sandy soil. Thus, if we want potatoes to form in the soil, and not have the plant shoot up into seed formation, we would rather have a sandy soil.

Steiner encourages the heightening of our observational powers. The lime forces work horizontally, as seen in the sedimentary layers of limestone, and the silica forces work vertically, as seen in the steep peaks of mountain ranges. He says we can trace the process quite exactly.

It is important to be able to distinguish between silica and lime forces. Steiner explains that in ancient times humanity created the different kinds of crops from primitive varieties by this kind of knowledge and instinctive wisdom. We must rediscover this wisdom and gain new knowledge of how these things work. As an example, Steiner says that people today do not know that silica receives light into the earth and makes it effective, while the lime and humus in the soil work in darkness.

Just as Steiner describes the farm organism as an upside-down person, we must also consider how the animals fit in. Different places on the Earth have their own specific types of animals. If, on a given farm, you have the right amount of animals to eat the farm's plants, the farm will have the right amount of manure.

The animals live in the belly of the farm organism, and we can see the influences of the planets in the forms of animals. From the animal's nose toward its heart, the distant planet forces are at work. In the heart itself the sun is at work, and from the tail toward the heart the inner planets show their influence. You have the true contrast of the sun and distant planets in the form and figure of the animal's head, and the moon and inner planets in the form and figure of the animal's rear. Knowing this, you will be able to discover a definite relationship between the manure of different animals and the needs of the earth where those animals graze. The animals that eat the farm's plants will in turn provide manure based on this fodder. Consequently, they will provide the very manure most suited for the soil that grew those plants. The farm is healthy inasmuch as it provides its own manure from its own stock. Studying skeletons of mammals in a museum can help us learn to read the forms of animals. As we learn to read nature's language of forms, we will perceive all that the self-contained farm organism needs.

Lecture Three: June 11, 1924

How do the forces Steiner has spoken of work in the farm through substances of the earth? He begins by considering nitrogen, whose significance in all farm production is generally recognized. But the essence of its activity has fallen into confusion. We must investigate the wider activities of nitrogen in the universe as a whole. Nitrogen, in living nature, has four sisters that combine with her to form protein (as was discussed in chapter 5). These sisters are carbon, oxygen, hydrogen, and sulfur.

Sulfur, a trace element compared with the others, acts as the mediator between the formative power of the spiritual and the physical. Sulfur and phosphorus got their ancient names as "light-bearers" because they carry light into matter. What chemists know of these substances in their laboratories neglects their inner significance in the workings of nature, much like what you learn of somebody from their photograph differs from what you come to know by meeting and getting to know them in person.

Carbon is the bearer of all living forms in nature. It is in all things that are alive or were alive. Through photosynthesis, lifeless carbon dioxide from the air becomes part of the living plant. Carbon provides the framework for the other sisters to move through the world and has the capacity to create the most varied and sublime forms. Life on Earth is based on carbon.

Every living thing is permeated by oxygen. Oxygen is alive inside of plants, animals, and humans, and is alive when it is in the soil. Science can only see dead oxygen, the result of considering it only from a physical standpoint. This is the dead oxygen in the air. Oxygen depends on nitrogen to find its way along the paths mapped out by the carbon framework.

Nitrogen has an immense power of attraction for this carbon framework, but carbon can draw nitrogen away from its other functions in the soil. Inert nitrogen is a corpse in the air, accounting for 78 percent of the atmosphere. It only comes alive in living beings. Often, there is more weight of animals, many microscopic, in the soil than above ground. Nitrogen becomes alive inside the earth, as does oxygen. But nitrogen is extremely sensitive, and this is of the greatest importance for agriculture. Nitrogen provides the basis for conscious awareness (intelligence) of the distant forces that work themselves out in the lives of plants, animals, and the soil. Nitrogen provides the intelligence that guides the oxygen (life) into the carbon (form). Steiner says that you can penetrate into the intimate life of nature if you can see the nitrogen everywhere, moving about like flowing, fluctuating feelings. We shall find the treatment of nitrogen infinitely important for the life of plants. In agriculture, above all, no nitrogen salts should be used.

Hydrogen is the first element in the periodic table, and the smallest. It is also by far the most plentiful element in the universe. There must be a constant interchange of substance between the Earth and the universe. In living structures, the spiritual becomes physical with the combination of these five elements. All that is living in the physical forms upon the Earth must be led back again into the great universe. It is not the spirit that vanishes, but what the spirit, carried by sulfur, has built into the carbon framework, which has drawn life to itself from oxygen with the aid of nitrogen. This must be able to vanish into the universe again. With hydrogen, the physical flows outward and is carried away, once more by sulfur.

Hydrogen carries the spirit and, using sulfur, connects with carbon, which provides the physical form. Oxygen carries life and nitrogen carries sentience. When an organism dies, hydrogen, again using sulfur, carries out into the far spaces of the universe all that was formed, alive, and sentient. So, we have these five substances, each inwardly related to a specific function.

Steiner then tells us what is really occurring when a person meditates. In meditating you always retain in yourself a little more carbon dioxide than normal and do not drive all of it back to the nitrogen in the air. Steiner gives the analogy of the experience of knocking your head on a table. If it's a hard hit, you will be overwhelmed by the pain—it's the only thing you'll be aware of. But if you rub your head on a table gently, you'll be conscious of both your head and the texture of the table. This is like meditation. By and by you grow into a "conscious living experience" of the nitrogen in the air around your body. All becomes knowledge and perception. By meditating, farmers make themselves receptive to the revelations of nitrogen and will farm accordingly. "All kinds of secrets that prevail in a farm and farmyard emerge." Even uneducated farmers can be meditators as they walk their fields. They acquire "a method of spiritual perception."

* * *

These five substances, provided by the atmosphere and bonded in the structure of proteins, are also bound to other substances and are not independent. There are only two ways they become independent. Hydrogen carries these fundamental substances out to the universal chaos or hydrogen drives them into the protein of seed formation. In seeds there is chaos, and in the "far circumferences" there is chaos.

For anything to grow, these five sisters need other substances, and these come from the earth. When carbon comes from the air into a plant and then into an animal or human, it must build on a content and framework not only of lime, but also of silica. The atmosphere, with its nitrogen and oxygen, must find a way into the soil, and one way it does this is through leguminous plants.

A bacterium that grows symbiotically on the roots of legumes can access nitrogen from the air in the soil. The nodules thus formed are visible and leave the soil richer in nitrogen, which is why legumes are essential in crop rotations. Steiner calls this a kind of nitrogen breathing, as if legumes were the lungs of the earth organism. Limestone, which drives this in-breathing of nitrogen, is always hungry, wanting to draw everything to itself, especially nitrogen. All that limestone craves lives in the plant nature. This must be balanced by an "uncraving principle" that desires nothing for itself, which is the insoluble silica nature. Like our sense organs that perceive their surroundings but not themselves, silica perceives the surrounding universe. Limestone provides the universal craving, and clay mediates between the two.

Lime underground disturbs carbon, so carbon, the form creator in plants, allies itself with silica to overcome the limestone nature. The plant lives in the midst of this process, with the limestone clutching below and the silica leading growth upward. But in the midst, giving rise to all the plant forms, carbon orders all these things. Humus is important because that's where the carbon is. We are beginning to see the right way to bring the nitrogen nature into the plants on our farms.

Lecture Four: June 12, 1924

Steiner begins lecture four with the observation that the world cannot be judged from conclusions drawn by investigating only at the microscopic level. For example, science had recently corrected itself regarding recommendations of daily protein consumption, from over 4 ounces a day to under 2. It is understandable that a science that only recognizes coarse material forces and substances will have to keep correcting itself: Such science considers soil fertility and human nutrition by focusing just on the nutrients and ignoring the dynamics. The greater part of what plants and humans ingest is there to give them living forces.

To help gain insight into the working of substances and forces, let's consider a tree and compare it to a mound of humus-rich earth. Mounding up earth gives it more vitality and makes it easier to permeate with humus, much like we do in making compost piles or raised beds. This earthly matter, permeated by humus, is on its way to becoming a plant, but as a mound it doesn't go as far as the tree.

Life continues from plant roots into the surrounding soil by the underground network of fungal hyphae, whose length boggles the imagination. The reason for fertilizing is to enliven the soil so this underground microbial network can help the plant get what it needs. We must know this in order to gain a personal relationship with our farming work, especially when working with manure and compost.

We can assess the vitality of living organisms with our sense of smell. Any living thing, from a cell to a whale, always has an inner and outer side, separated by some kind of skin. The insides will concentrate and organize the smells, while the outside lets them go. Plants tend to absorb odors. Steiner enthusiastically tells us to perceive the helpful effects of a fragrant meadow of aromatic plants.

Manuring is essentially communicating livingness to the soil, yet not only livingness. We must also enable the nitrogen to spread out into the soil. With its attraction for carbon, nitrogen can carry the life property to the plant's roots. Lime helps with this, too. Because water-soluble fertilizers lack carbon, they can't bring life to the earth. Although they can stimulate plant growth, they are harmful to the soil's life.

In compost we have a means of kindling life in the earth itself. Not only does compost have the living oxygen principle, but the carbon materials absorb the sentient nitrogen principle, and this is most important. Manures already have the digestive, alkaline lime principle working well enough. But for composts made without animal manure, a small amount of quicklime, a couple pounds per ton, will help absorb the oxygen without volatilizing the nitrogen as ammonia. Too much lime causes volatilization. The lime absorbs not only the nitrogen but also the oxygen, so fertilizing with this compost communicates the nitrogen directly to the soil without going through the soluble nitrate phase. When we fertilize meadows and pastures with such compost, the hay and fodder will be much better for our animals.

Keep in mind that we depend on our feelings, which will develop once we perceive the whole nature of the process. We can also develop the necessary personal relationship with our sense of smell. We want the compost heap to smell as little as possible, so we alternate thin layers of manure and soil with layers of organic materials such as peat, old hay, or rotten wood chips. The nitrogen, which otherwise might evaporate, is now held by the carbon in these materials.

Organic entities, like a compost pile or a cow, have streams of forces pouring both outward and inward. But the horns and hooves of cows don't allow the cow's internal forces out. Instead, the horns and hooves ray them back into the cow's digestive tract. The horn is especially well adapted to ray back the living forces of oxygen and the sentient forces of nitrogen. Antlers are altogether different and discharge streams and currents of forces outward. Animals with antlers are nervous and quick, while horned animals such as cows are calmer and slower.

The grass a cow eats is not completely excreted for many days. During this time a host of microbial activities take place in her belly, and the manure is permeated by forces that carry nitrogen and oxygen. Forces in the digestive tract are like those in a plant. Manure has a life-giving and sentient influence upon the soil, and upon the earthly element, not just the watery element like artificial fertilizers. The increase of microbes indicates the manure is in good condition to support life, but inoculating manure with bacteria is not how to improve it.

Instead, take cow manure, stuff it into a cow's horn, and bury it in good soil in the fall after the equinox. This preserves the oxygen and nitrogen forces of the manure just as the horn did when it was on the cow. Over the winter the manure inside the horn is enlivened with these life and sentient forces, which attract and concentrate the plant-like living oxygen and animal-like sensitive nitrogen forces within the soil. The manure in the horn transforms into a humus-rich fertilizing substance with a highly concentrated, life-giving manuring force.

Life forces beget more life, as organization flows from lower to higher concentration. Creating something rich in life and intelligence breeds more of these qualities. Throughout the winter the content of the horn becomes inwardly alive, because winter is when the earth is most inwardly alive. By contrast, the life of the earth flows outward in summer into the growth and activity of all the plants and animals.

The resulting horn manure, dug up at the end of May, is stirred quickly in a pail of slightly warmed water, changing the direction after each deep vortex is formed. As it seethes around in the opposite direction, bubbles show that air is being incorporated. After an hour of stirring in alternating directions, a thorough penetration is achieved and a delicate aroma arises. This personal relationship is important, and you can easily develop it. The contents of the pail are then sprinkled over the land. Combine this with the ordinary use of compost and manure and you will soon see how great fertility can result. The use of horn manure can be followed up at once by another preparation.

Grind quartz crystals into a fine, floury powder and add water to make a mush. Steiner describes the product as being like "a very thin dough." Fill a cow's horn with this, bury it over the summer in the soil, and dig it out in late autumn. In this case you need far less of the dough—a fragment the size of a pea. Like the other, you have to stir this in water for an hour. You can use this silica water to sprinkle plants externally, and it will prove most beneficial with vegetables. You will soon see how the cow horn manure drives from below upward, while the cow horn silica draws from above.

At the end of the lecture, Steiner notes that although agricultural research investigates ways to increase production for financial profit, the most important point is not quantity but quality. Steiner believed the method he described would result in the very best food for humans and animals. "The end in view is the best possible sustenance of human nature."

Question and answer sessions took place after lectures four, five, six, and eight. I have included a few excerpts from these discussions.

Discussion: June 12, 1924

A listener asks whether one can use a mechanical stirrer instead of stirring by hand, and Steiner replies that there is a difference. When you stir manually, he says, "All the delicate movements of your hand will come into the stirring. Even the feelings you have may then come into it." With enthusiasm, great effects can be called forth. As a result of the concentration and subsequent dilutions, it is no longer the substances as such, but the dynamic, radiant activity that is working.

It is best to get fresh horns from medium-aged cows that have been living near your farm for three or more years. They should be 12 to 16 inches long and can be used for three or four years. Don't disinfect the horns. People go too far in disinfecting things.

To store the horn manure, make a box upholstered with a cushion of peat moss on all sides so the strong inner concentration will be preserved. But once the manure is stirred in water it should be used within a few hours. The microbes inherent in the horn manure are propagated during the incorporation of air and water but cannot live long after the stirring stops. It will do the horn silica preparation good to be stored where the sun can shine on it, but it also needs to be used right after stirring.

When asked if the fine spraying weakens the forces, Steiner says no. Unless you drive them away yourself, you need not be afraid about spiritual things running away from you nearly as much as with material things.

To questions regarding stronger concentrations for more rampant growth, Steiner recommends staying with the amounts he has given. Quantities are now accepted as $\frac{1}{3}$ cup of horn manure in 3 gallons of water for 1 acre, and $\frac{1}{2}$ teaspoon of horn silica in 3 gallons of water for 1 acre. The former is applied in the evening before sowing so as to influence the soil itself, while the latter is applied in the morning on the growing plants.

In response to the broad question—"Can anyone you choose do the work?"—Steiner replies that some people have a "green thumb" and others don't. As he describes in lecture three, "Such things will come about simply as a result of the human being practising meditation. . . . When you meditate you live quite differently with the nitrogen which contains the Imaginations. You thereby put yourself in a position which will enable all these things to be effective."

There were times in the past when people knew that by certain definite practices they could make themselves fit to tend the growth of plants. Delicate and subtle influences are lost when you are constantly living and moving among people who take no notice of such things. Old folk proverbs contained a peasant's philosophy, a subtle wisdom about such timing. Steiner regrets the loss of the cultural philosophy common in his youth, when you could learn far more from the peasants than in the university.

Someone asks if they should determine the number of cow horns their farm needs purely by feeling. Steiner doesn't advise this, saying we must be sensible. Test it thoroughly according to your feeling, but then translate the results into figures. Bear in mind that the judgments of the world are tending toward calculable amounts, so compromise in this respect as much as possible.

Steiner's final answer during this discussion session is extremely important. He points to relearning how farmers fertilized before artificial fertilizers were available. He reminds us that this new method of cow horn manuring is not a substitute for ordinary manuring. The horn manure enhances the effect of the regular manuring, which should continue as was done before.

Lecture Five: June 13, 1924

The *Agriculture Course* is 174 pages long, so page 87 is the halfway point. Here Steiner reiterates, "The preparation I indicated yesterday for the improvement of manure was intended, of course, simply as an improvement, as an enhancement. Needless to say, you will go on manuring as before."

In Steiner's youth, farmers raised livestock and practiced strict crop rotations with legumes, grasses, grains, and other crops. Soil moisture and microbes were conserved by careful attention to liming, composting, and the art of plowing. Farmers recognized that the soil is a continuation of the plants growing in it. Science no longer perceives this common life of the soil and all plant growth, nor how it continues in the manure.

Steiner disagrees with science's insistence that the elements found in plants are the only ones that are essential for proper growth and that those elements should be added to soil in a water-soluble form. This ignores the role of silica, lead, arsenic, mercury, and sodium, which are regarded as mere stimulants. But according to Steiner, these elements are the most important of all and are provided freely in rainwater. So we don't have to add them to the soil like we do potash, phosphate, and limestone.

Human activities can keep the soil from receiving silica and the other "mere stimulants." They become unavailable for plant growth when soil is compacted, or when we use artificial fertilizers and till randomly and carelessly. When this is the case, additions to manure or compost will help. It is not substances but living forces that should be added. It is easy to use minute quantities in the right way to set free radiant forces in manure and compost.

Steiner then mentions and describes preparations to add in minute doses to the manure and compost to bring the nitrogen nature into the world of plants. "To-day therefore—more as a general indication—I shall mention a few more things in the same direction: preparations to add to the manure in minute doses. . . ." This implies that these are general ways of doing things, and there are other ways than the ones he will be indicating. This process will be accomplished homeopathically, with small amounts of herbs that have been specially treated.

He notes that replacements can be found if certain things are difficult to find. We must provide the right way for carbon, oxygen, nitrogen, hydrogen, and sulfur to come together with other substances, notably with potash. It is very important that the potash assimilates itself rightly with the protein in the plant.

To do this we take yarrow, a miraculous creation that shows how to bring sulfur in the right relation to the substances of the plant, particularly the potash. Steiner then explains how to make this preparation (as described in chapter 8). After digging it up, we are told to put a few grams in a manure or compost pile, which can be the size of a house, and the radiant force will influence the whole mass. This endows the pile with the power to enliven the soil, enabling it to receive silica, lead, arsenic, mercury, and sodium from the atmosphere. The minute amounts of sulfur in yarrow, combined in a model way with potash, enable the yarrow to ray out its influences through large masses and great distances.

Through its antlers, a stag is intimately aware of the periphery in its environment and of what is going on all around it. In the stag's bladder we have the necessary forces to give yarrow the power to enhance its own ability to combine sulfur with the other substances. It is important to observe that we have a fundamental way of improving manure while remaining within the realm of life and not resorting to inorganic chemistry.

The purpose of the next preparation is to give so much life to our compost and manure that it is able to transmit life to the soil. The manure needs to become humus-like so it can bind together the substances necessary for plant growth. In addition to potash, we need calcium compounds. While yarrow assimilates potash with the help of sulfur, another plant also assimilates calcium, and that's chamomile. You can follow the process that these herbs undergo when taken medicinally. Yarrow is used medicinally for bladder infections, and chamomile for intestinal problems, so Steiner's choice of pairing these herbs and organs together isn't so far-fetched for an herbalist.

After clipping off the yellow and white flower heads of chamomile, we stuff them into cow intestines. We bury these sausages in humus-rich soil. Choose a spot where snow will remain and leave them through the winter. Add a small amount to the manure, as you did with the yarrow preparation, and the manure will develop a more stable nitrogen content. It will kindle life in the soil and will lead to much more healthy plants if you fertilize this way.

Stinging nettle is the greatest benefactor for plant growth in general, and it would be hard to find a replacement. Besides sulfur, stinging nettle carries potassium, calcium, and iron in its radiations. These iron radiations are almost as beneficial to nature as they are in our blood. To improve the manure still more, bury a mass of the leaves and stems of stinging nettle, surrounded by peat moss, in the soil. Dig it up after it has spent the winter and the following summer underground. Use small amounts like the other preparations and the manure becomes inwardly sensitive and intelligent, and will not lose nitrogen. It's like permeating the soil with reason and intelligence.

The prevention of diseases takes a more general course in plants than in animals. A rational improvement of manure, using the following method, can remove many plant diseases. The manure must bring calcium into the soil, but this calcium must remain within life, so ordinary lime is of no use. The bark of a white oak tree contains plenty of calcium in an ideal form. The outer bark is an intermediate product between plant and humus, as described in lecture four when comparing a tree to a mound filled with humus. This calcium has the property of damping down life and restoring order when growth gets too rampant, which is when disease appears. Calcium's unique structure in oak bark helps this process without creating shocks in the plant. We grind the bark up and pack it into the brain cavity of a cow or other domestic animal's skull and cover the spinal cord hole with a bone. This time we bury it in a marshy damp place with decaying vegetation throughout fall and winter. When added to the manure pile it lends forces that arrest and reverse nitrification and combat plant diseases.

Next, we need something to attract silica from the whole, distant environment. We don't notice that the soil gradually loses silica in the course of time, nor do we notice silica's great significance for the growth of plants. Scientists in Steiner's time had just started accepting the idea of the transmutation of elements, as in radioactivity. If the processes taking place around us all the time were not so utterly unknown, the things Steiner explains would be more believable. A modern agricultural student might now argue that Steiner hasn't told us how to improve the nitrogen content of the manure. Though it is quite a foreign concept, the improvement in nitrogen content occurs due to the alchemy in the use of yarrow, chamomile, and stinging nettle.

If potash is working properly through herbs, animal organs, and the resulting biology, this hidden alchemy transmutes the potash into nitrogen. It even transforms the limestone into nitrogen. Calcium helps the legumes grow, and their symbiotic bacteria incorporate atmospheric nitrogen into the soil biology. These transmutations require microscopic organisms in the soil.

Steiner notes that the ratio of nitrogen and oxygen in the air, approximately four to one, is similar to the relationship between lime and hydrogen in the life processes. He then suggests that under the influences of hydrogen and carbon, lime and potash are constantly being transmuted at length into nitrogen, which is of the greatest benefit for growing plants. This is the nitrogen I want.

My understanding is that the elements calcium and potash become part of the bodies of microbes. As they work their way up the food chain, nitrogen from the air in the soil also becomes part of these microbes. As Steiner points out, these processes are not completely understood and can therefore be called mysteries. After minerals are incorporated into a microbe's body, nitrogen gets coaxed into the body with the help of carbon and oxygen. It then becomes part of amino acids, the ideal form of nitrogen for plant growth. All this happens under the influences of hydrogen, or you could say under the influences of the entire heavens, because hydrogen is by far the most common element in the universe.

Silicon is also transmuted in living organisms, such as through the forming and dissolving of an insect's exoskeleton. In plants there must arise a clear and visible interaction between the silica and the potassium, but not the calcium. Lime and silica are polar opposites. We can fertilize the soil so it aids in this relationship by finding a plant whose own relationship between silica and potassium can impart this power to the manure. Steiner then introduces dandelion as "a kind of messenger from Heaven" that mediates between the tiny amounts of silica, distributed in the atmosphere and beyond, and what is needed in the soil. After dandelions are prepared (as described in chapter 8), they're added to manure or compost. It gives the soil the ability to attract just as much silica from the atmosphere as the plants need to make them sensitive to what is at work in their environment. Plants grow better if they can sense all that is in the soil and above them and enlist these things to help them grow. They do this with the aid of their symbiotic microbial network.

Although you can artificially fertilize and limit the environment the plant needs, it is not good to do so. Steiner proposes we create better fertilizers by adding yarrow, chamomile, stinging nettle, white oak bark, and dandelion, or suitable substitutes, to our compost and manure, instead of using all manners of chemicals. Treat the soil with these preparations and the plant can draw what it needs from a wide circumference, including adjacent meadows and forests.

As a final step, before using the compost or manure, press out the flowers of valerian, dilute the extract, stir briefly as with the horn manure, and then sprinkle this over the compost or manure heap. Phosphorus, which tends to be locked up, can be made available by the biological activity the valerian flower juice stimulates. With the help of these six ingredients, excellent fertilizers can be produced, whether from liquid manures, farmyard manures, or composted organic materials.

Discussion: June 13, 1924

In response to a question about whether manure should be covered in rainy areas, Steiner replies that manure does not need a roof over it. Although too much rain would wash it out, in moderation rainwater is good for manure. A skin of hay or peat is a good idea. In response to a question regarding a new method of loosely piling manure to generate warmth and become odorless, Steiner reminds us that success with new methods may not always be practical in the long term. A new medicine can work wonders the first time you take it, but its curative effects may diminish if you keep taking it. The spontaneous generation of warmth in manure piles is exceedingly good for the manure, and becoming odorless indicates the method is beneficial. The pile does not need to be in a pit or on a hill, just built up from ground level. Don't pave underneath it but add clay if the soil is sandy or sand if the soil is clayey.

One of the farmers says they'd been eradicating yarrow and dandelion, thinking they were bad for cattle. Steiner suggests watching carefully, noting that animals instinctively know what is not good for them to eat. We only need to use a small amount of the herbs recommended. When we want to stimulate something alive, we often use what we would not otherwise use. Many medicines are poisonous in large quantities.

More questions and discussion about preparations and nutrition follow, including these points:

- The preparations are not buried together, but in separate spots on the farm, in the fertile layer of topsoil. They can be overgrown with plants. Once you have dug them up, place them separately in holes about 18 inches deep in the pile and close the manure up

around them. It doesn't matter how long the preparations are kept with the manure. The purpose of
the preparation is to radiate forces throughout the
pile. Placing them not too near the surface or each
other will keep the radiations from going outside or
interfering with one another.

• The yarrow, chamomile, nettle, and dandelion preparations are buried in fertile soil, not subsoil, and
frost will do no harm. Frost exposes the earth to
distant influences.

• The outermost bark of a white oak is taken from a
live tree.

• In the absorption of food, the forces developed by
the body are the essential thing. The animal must
receive the proper food to be able to develop sufficient forces to absorb what it needs from the atmosphere. Too much food can shorten an animal's life.
Like us, animals need to eat the correct amount, not
too much or too little.

Lecture Six: June 14, 1924

Steiner begins by giving examples that, taken as a
starting point for experimentation, will lead further
into ideas regarding harmful plants and animal pests.
For instance, a weed might be a useful plant, although
growing where we don't want it. To deal with plants
growing out of place we are first reminded of the
distinction between the forces of growth and reproduction working downward into the soil, and the forces of
nutrition working upward into the air. The former are
influenced by lime, and the latter by silica.

Both the soil and agricultural traditions have become exhausted, though sometimes simple peasant folk can lend a hand. For example, science is at a loss to find a remedy that prevents aphids on grape shoots. Aphids are beneficial in the soil via their interaction with ants and fungi, but detrimental when up in the plant canopy. The solution needs to recognize that there are soil forces up in the plant that should be down in the soil. Peasants were aware of these forces.

Weeds are often medicinal plants, influenced by the moon. We only see sunlight reflected by the moon, but there are other forces as well. Lunar forces strengthen the soil, intensifying its normal vitality to the point where plants can reproduce themselves. Reproduction is simply an enhanced growth process, and lunar forces enhance plant growth to the point of reproduction.

Weeds become particularly troublesome in wet years when we can't get in the fields to cultivate. However, we can learn how to weaken lunar influences and make weeds reluctant to grow. Gather weed seeds and burn them in a wood fire. Just as water brings about fertility, fire destroys it. The ash concentrates the force opposite of what water attracts. By scattering this small amount of ash over the fields, the weeds will grow less rampantly. We may have to repeat this every year for four years, as these influences occur in four-year cycles.

People used to know these things instinctively and could rid their fields of weeds by such practices. Steiner says that he can only offer suggestions, but they are quite practical. He is sure tests can verify what he proposes, but if he had a farm he would apply the methods at once. One can know things inwardly, and what is inherently true will later be confirmed.

We must be less general when dealing with animal pests than with weeds. Let's take the field mouse as an example for our experiments. Poisons of phosphorus and strychnine were used to kill mice. State regulations in Steiner's time were made requiring neighborhood farms to do this too, or the mice would just come back from there. What we do to control animal pests would also be better if the neighbors followed suit, but insight into the method is required for neighbors to see the sense in doing it. Intelligent insight works much better than police regulations.

When we move from plants to animals, we need ideas that consider not just our solar system but the fixed stars, notably the constellations of the zodiac. The word "zodiac" means "animal circle." Although lunar influences are enough to bring reproductive forces to plants, animals also need forces from Venus for reproduction. As with weeds, the reproductive force is weakened by fire.

Catch a fairly young mouse, skin it, and burn the skin when Venus is in the constellation of Scorpio. The burnt mouse skin inhibits field mouse reproduction. Sprinkle this ash over the fields and the mice will avoid the area. Again, only a small amount is needed. We can get much pleasure in farming this way, reckoning with the influences of the stars without becoming superstitious. However, to deal with insect pests, we need to understand that they are subject to different starry influences than the higher animals.

For example, a nematode infesting sugar beets makes the roots swell and the leaves look limp in the morning. Steiner reiterates that leaves absorb atmospheric influences and that roots absorb influences from the soil. Nematodes appear when the leaf and air forces are pressed downward into the soil. All living creatures can only live within certain limits and with certain forces. Aphids can only live above ground when forces normally in the soil are up in the plant's leaves. Root-knot nematodes live when forces normally in the air are found in the soil.

Burn the insect when the sun is in the constellation Taurus, which is the opposite sign from Scorpio. Sunlight reaches the Earth's surface at lower or higher angles when the sun is in different constellations, so different forces are involved. Spread the ash on the infested fields and the pest will shun life there, especially if you do this each year for four years.

It is important to relate to the soil this way. Burning seeds annihilates their power of fertility. Burning insects and animal skins in the appropriate constellations annihilates the power of fertility in these animals.

To deal with plant diseases we need to notice that these arise when the forces of life are not strong enough to mitigate the effects of overly strong lunar forces. This happens especially when a wet winter is followed by a wet spring. The earth becomes too strongly alive and pushes growth too fast. Then the aboveground part of the plant becomes a lush medium for parasitic growths such as mildews, blights, and rusts. The overly intense lunar influence from excessive moisture interferes with the forces of forming and ripening seeds coming from the outer planets. Several significant relationships are involved. Wet conditions in the soil overwhelm the lunar growth forces, producing nitrates in the wet soil that can cause plant diseases. When the soil is dry, the life-giving influences of the moon are not as strong. Although the lunar forces are necessary for seed formation, they must not be too strong.

When we perceive what prevents the soil from absorbing excessive water, we think of sand and good drainage. Sand is silica, which has a drying effect. Horsetail is a plant rich in silica that we can use in small quantities. Prepare a decoction of dried horsetail by simmering it for thirty minutes. Dilute this and sprinkle it over the fields. This is not so much a healing process as it is the opposite process to the overly intense lunar forces. Silica is the antidote for too much water, as it absorbs excess moisture.

Real science arises when we learn how to control the overall forces at work. Life cannot be understood in isolation from the whole. Modern science's analytical, microscopic way of studying nature must give way to an understanding of the macroscopic, of the entire universe from which life proceeds. Steiner tells us that "nature is a great totality; forces are working from everywhere."

Discussion: June 14, 1924

Asked about the insect "pepper," Steiner replies that both the larvae and the adult can be used, although it might shift the constellation that the sun should be aligned with. The proper constellation moves from Aquarius to Cancer as the winged insect passes to the larval stage. We will have to test burning the different kinds of insects at different times in the annual cycle of the sun passing in front of the various zodiac signs and constellations. When asked about the right time to burn animal skins, Steiner says that Venus should be behind the sun and in front of Scorpio.

The next question asks whether the specially prepared manure, besides being given to turnips and garden crops, is also important for grains. Steiner says to continue the methods that have proven effective, and just supplement them with these new methods. The influences will be lesser with sheep or pig manure than with cattle manure.

To a question about inorganic fertilizers, Steiner responds that mineral manuring, as with NPK fertilizers, must cease because the products from fields thus treated lose their nutritional value. With the methods he has given we'll be able to gradually stop using chemical fertilizers. These methods will be much cheaper, and the soluble fertilizers will go out of use. Steiner suggests we consider how everything is being mechanized and mineralized, yet minerals only work as they do in nature. Do not put lifeless artificial fertilizer into a living soil.

He goes on to say that in a recent discussion on beekeeping, a modern beekeeper was especially keen on the commercial breeding of queens. Queens were being sold, instead of being bred within the single hives. Steiner says he had to reply that modern opinions are based on far too short a period of time. He says that if not in thirty or forty years, then certainly in forty to fifty years' time beekeeping will thereby be ruined.

Lecture Seven: June 15, 1924

Although scientists frequently study things in nature isolated from their whole environment, Steiner begins the seventh lecture by emphasizing that all things in nature are in mutual interaction. In times past, farmers were thoroughly familiar with the interactions of minerals, plants, and animals through an instinctive insight that has now been lost. Today, scientists only study the coarse interactions, not the effects of the finer, subtler interactions.

We must observe these more intimate relationships, which are constantly taking place, when we are dealing with the lives of plants and animals together on the farm. Besides our crops and livestock, we must observe with intelligence the colorful world of insects, hovering around plants, and learn to look with understanding at the birds. Modern society has caused a reduction of bird populations in certain areas and doesn't realize how greatly farming and forestry are affected, Steiner adds.

When we look at and consider a tree, it's noticeably quite different than an herbaceous plant. We can only include the flowers, leaves, and shoots growing out of the tree branches annually as the plant nature. The rest of the tree, the older branches and trunk, is more like a mound of soil upon which the plant nature is growing. Right underneath the bark is the cambium layer that connects the new growth to the soil, like a long thin root for those "plants" growing on the branches.

To help us understand what a root actually is, Steiner makes this comparison. If a bunch of plants are growing close together, their roots would intertwine and merge with one another. As you can imagine, such a complex of roots would not remain a mere tangle of roots winding around each other, but would grow organized into a single entity. Because the saps and fluids would flow into one another, you wouldn't be able to distinguish where the different roots begin or end. Something like a common root being would arise for these plants.

Nowadays we call the fungal roots connecting plants together underground a common mycorrhizal network. Communication happens between plants through this network. A plant infested with aphids, for example, will within a day transfer a signal that elicits a nearby plant to release odors that attract aphid-eating insects, thus protecting itself. There is no hard and fast line between the life in a plant root and the life in the soil because of this common mycorrhizal network.

The cambium layer is where cells divide and make new cells. This is the living, growing layer underneath the bark. The tree's trunk is like soil that has bulged upward into the air. Having grown outward, the tree needs more inwardness, more intensity of life than an herbaceous plant does. Trees grow high up in the air where the atmosphere is different than directly over the soil where other plants grow. There is more of the oxygen process happening around herbaceous plants and more of the nitrogen process up in the treetops. A deep dark forest feels very different than a sunny meadow.

Steiner recommends we acquire a certain sensitivity to the aromas of herbaceous plants compared to trees. The scents wafting from blooming fruit trees and forests are rich in nitrogen, while the other plants smell more earthy. This is because the life and oxygen forces are depleted underneath where trees are growing. Tree roots last a long time and become more mineralized. When we clearly envision this, we'll notice that insects live in the air, and their larvae belong underground.

* * *

Also in the soil are the wonderful earthworms. Study how they live together with the soil. Their castings are a perfect example of a humus substance containing what the plant roots need in an available but colloidal form. This allows plants access to nutrients, but since these nutrients are not in a water-soluble form they don't leach out when it rains. Steiner recommends that farmers take special care of the soil to encourage the extremely beneficial activity of earthworms.

There is a remote similarity between insects and birds. Insects flutter about primarily in the meadows and leave the treetops for the birds to fly around in. Both help to distribute nitrogen forces wherever they are needed. Winged animals are unthinkable without plants and vice versa. Farmers should have some understanding of the care of birds and insects, for in nature everything is connected.

Steiner says that these things are important for true insight, and we should clearly place them before our souls. What the flora and fauna of forests provide for the surrounding area has to be accomplished by quite a few other things in un-wooded areas. The growth of soil is subject to different laws in areas where forest, field, and meadow alternate than it is in wide stretches of un-wooded country. Observe and then contemplate these things clearly. We should have the insight to preserve forests in districts where they existed before human intervention. Forests are good for the surrounding farmland. We should have the heart to increase woods when we see vegetation stunted or make clearings in the forest if we notice plants growing rampantly and not producing seeds. The regulation of woods and forest is an essential part of agriculture.

The world of earthworms and other soil life is related to limestone, the mineral nature, while the world of insects and birds is related to silica. Things like this were recognized in olden times by instinct. Earthworms require lime in the soil, and birds need silica-rich conifers around them. Realizing this, another kinship emerges, the inner kinship of mammals to shrubs and bushes. It helps to plant shrubs in our landscape because mammals love them.

Another intimate relationship is the one of mushrooms or fungi to bacteria and harmful parasites. The meadows on our farms should be well planted with mushrooms and toadstools. A meadow rich in beneficial fungal activity helps prevent harmful fungi and bacteria from affecting the plants we grow on the rest of our farm. It is not economical to rid our farms of these things, hoping to increase crops. We will get worse quality by increasing tilled acreage at the cost of removing these other things. Farming is intimately connected with nature, and to engage in it you must have insight into these mutual relationships of nature's husbandry.

We have to ponder the relationship between plants and animals in order to learn how to feed our livestock. By observing what is in a mammal's environment we can perceive what is happening. A mammal receives and assimilates air and warmth in its nervous system and also in part of its respiratory system. It's a creature that lives directly in the air, breathing in oxygen, and creating warmth in its blood flow. But animals cannot assimilate water and earth directly. They must drink and eat, so they must have a digestive system to do this.

On the other hand, a plant doesn't need a digestive system. It has an immediate relationship to water and earth, just as the animal has with air and warmth. The plant lives directly with water and soil. Plants do not take in air and warmth internally like animals do with water and earth. Plants give off oxygen and a slight degree of warmth. The plant's life processes are the inverse of those of the animal. The expelling of air and warmth has the same importance for the plant as drinking and eating do for the animal. The plant lives by giving. Everything in nature lives by give and take. The symbiotic relationship between plants and animals is harmoniously ordered by their environment.

Orchards, shrubs, and forests are regulators to give the right form and development to the growth of plants throughout the whole farm. Earthworms and lower animals, in unison with the soil's lime content, act as regulators down in the soil. That is how we must regard the relation of our tilled fields to the orchards and the forests, with their insects and birds, and to the mushroom-rich meadows with the grazing livestock. "This is the very essence of good farming."

Lecture Eight: June 16, 1924

In his final lecture, Steiner again recommends that farmers develop insight to be able to act individually with intelligence on the practical hints he still has to offer.

Plants have their physical mineral body and also a living body that builds and maintains their physical presence. An awareness and intelligence hovers around a plant and connects with the plant when it produces edible nourishment in support of animal and human awareness and intelligence. Insight reveals whether a certain feed or food supports some process in animals or humans. After food is ingested, it isn't burned or combusted in the body, which is a process in lifeless nature. What takes place within the body is quite different. The process is altogether living and intelligent.

Our human system is threefold, consisting of the head pole, the limbs and metabolism on the opposite pole, and a well-defined rhythmical system in between, composed of the heart and lungs. Animals have the two outer poles as well, but the middle, rhythmic one is not as independent as it is in humans.

The substances of the head, nerves, and sense organs come from the soil through what is digested.

The substances that make up the limbs and metabolism, on the other hand, come from what is absorbed
out of the air and warmth above the soil.

The opposite is true of the forces. In the head, with
its nerve and sense system, are forces of a distant, universal origin. The rest of the body is subject to local,
earthly forces, such as gravity. In practice, you need to
feed a work animal so that it can absorb the substances
out of the air to strengthen its body. What is needed by
the head, rather than the muscles, must be gotten from
the actual fodder.

Distant and atmospheric substances cannot flow
easily into a dark stable. Animals should be out on pastures and have the opportunity to sense the surrounding world. Compare the confined animal, eating what
humans decide it needs, to one that is using its own
sense of smell to find food outside. The indoor animal
has inherited traits that conceal its lack of forces from
outside, but its descendants will show the lack. Thus,
future generations of animals will be weak and won't
be able to absorb what they need to be truly healthy.

In the head you have the brain, which in humans
serves as the basis for our ego, our self-consciousness.
The animal does not have self-consciousness. Its brain
is only on the way to forming an ego, but it is not
there. The substance of the brain comes from what
the animal has eaten. Some of the earthly substance
is digested and excreted, but some continues on to be
deposited in the brain. In humans, more of what's in
the belly reaches the brain than is the case with animals. So the animal's manure has more potential for

self-consciousness, whereas in humans that potential proceeds to the brain. Animal manure brings the potential for self-consciousness to the roots of plants. This potential connects the roots to the soil's microbes and minerals, so these plants can access all they need and in turn provide the best possible nourishment, not only for the microbes but for higher animals as well.

A farm is a living organism, with nitrogen forces developed in the fruit trees and forests, and oxygen forces in the meadows. As animals eat the pasture, they develop forces that are given over to their manure. These forces enter the soil with their manure and cause the plant roots to grow deep in the earth, following gravity. A farm is also an individuality, and you will gain the insight that your animals and plants should be kept within this mutual interplay. You hinder the way nature works by bringing in nitrogen fertilizers. Letting the farm's own animals fertilize the farm creates a perfect, self-contained cycle. Each farm, as a unique individual entity, requires the right kinds and amounts of animals. All we can do here is indicate general guiding lines. As far as possible we should make our farms able to sustain themselves without becoming fanatical about it. In life, with the way economics works, this may not be fully attainable, but a self-sufficient farm is something to aim for.

Observe a root and see how it absorbs the manure's forces. The plant is assisted by the forces in the manure as well as the salts in the earth. This becomes the food we eat and eventually nourishes our head and nervous system. A growing animal also needs this nourishment, so we feed them roots. Carrots are traditionally fed to calves. You also need a second kind of food to help bring the forces in the head to the rest of the body. It needs to be something that rays out in the plant. Linseed or fresh hay added to the carrots will make good feed for young growing cattle. Any combination of a root crop with a long, thin plant like hay will stimulate the head and assist what is needed to pass downward to fill the body. This is another example of the principle that we should observe the things themselves and learn what happens to them as they go from an animal into the soil as fertilizer, and also as they go from the plant into the animal as feed.

On the other hand, for good milk production we need to strengthen the middle part of the animal. We need the right cooperation between the forces streaming back from the head with the substances that pass forward from behind. We don't need roots in this case, or blossoms or fruit. For good milk production we want what is between these—the green vegetation. To further stimulate the development of milk we use plants that have their flowers and fruits close to their leaves. These are the legumes, most notably clovers. It takes a generation to see the results. The offspring of animals treated this way will make high-quality milk.

Another point worth noting is that the fruiting process can occur in other parts of the plant. Although for most plants we sow the seed of the fruit, with the potato we use their "eyes." The ability to propagate doesn't always follow a pollinated flower. It can be latent in another part of the plant, such as a bud, that can make roots. This is called asexual reproduction.

The fruiting process can be enhanced by the combustion process, such as drying hay in the sunshine. We heat and cook food because warmth plays a considerable part in the fruiting process and makes the food more digestible. We enhance the fruiting process of plants that quickly go to seed, such as legumes, by cooking, boiling, or simmering them.

To fatten animals, you feed them fruiting substances, possibly further enhanced by cooking or drying. Also, give them food that has a fruiting process enhanced by cultivation, like domesticated turnips or beets. When grown with human care, plants grow bigger than when in the wild. Oil plants are helpful to distribute what the animal absorbs. But we also need something of the root nature to allow the earthly substance to pass upward to the head. We have roots for the head; flowers, fruits, and seeds for the metabolic and limb system; and leaves for the rhythmic system in the middle. The last thing needed for the whole animal is a small amount of salt. Small quantities fulfill their purpose if the quality is right.

Steiner ends this lecture by saying he could go on giving many individual guiding lines, which are only foundations for many experiments. He advises waiting to publish experimental results until after the farmers confirm them, because it makes a great difference whether farmers are speaking from direct experience or non-farmers are speaking theoretically.

Discussion: June 16, 1924

In response to a question about using electricity to preserve fodder, Steiner replies that electric currents and radiation have a negative influence on human development. Consider how electricity is now being used above the Earth as radiant and as conducted electricity to carry the news as quickly as possible from one place to another. People living around electricity, notably radiant electricity, will no longer be able to understand the news that they receive so rapidly. The effect is a "dampening down of their intelligence."

Everything that enters an organism as food must undergo a complete change, and salt can help. Even warmth must be changed in the body. If it doesn't, we catch a cold.

A listener asks about the karmic effects of making the insect peppers, and if it matters what frame of mind a person is in. Steiner responds with a reminder to consider the whole way we must think. In lecture eight he points out how one must see from the outer appearance of the linseed or the carrot what kind of process it will undergo inside the animal. You would have to have evil intentions to cause harm, so common morality should also be fostered. Killing a live being, like a fish, is different than destroying fish spawn. Preventing the conditions for a host of beings to be born is different from killing things.

A question is asked about using human feces as fertilizer. Steiner responds that this should be used as little as possible, no more than what the people on the farm produce.

The final question asked is about green manuring. Steiner replies that the practice has its good sides and is useful for certain plants, especially if you want to encourage a strong effect on leaf growth.

In concluding the lectures, Steiner says he feels that besides the deep inner value of what took place during the agriculture course, which was real and useful work, they had also all enjoyed a "real farm festival."

The Essence of Good Farming

An early-morning stirring of horn silica is following another Memorial Day weekend of making preparations. An acre of blooming potatoes, a ripening garlic patch, and other spring crops are ready. My palm cups the soft, warm rainwater and spins it around, making a vortex almost to the bottom of the bucket. After twenty seconds I pull my hand out, admire the oscillating mandala, and then begin to stir in the opposite direction, forming a seething, bubbly chaos. An hour later it is done, and then the atmosphere above the fields fills with a fine mist of homeopathic silica.

Although the horn manure we dug up last weekend was excellent, I couldn't find the yarrow, the dandelion wasn't too impressive, and a skull had lost its oak bark. Oh well, try and try again. We stuffed five stag bladders with beautiful yarrow florets and filled two big tiles with stinging nettle leaves we had stripped off the stems. We hung the yarrow in the sun and buried the nettle in a deep pit surrounded by chopped-up leaves. A cow skull, fresh from the one we took to the butcher, was filled with white oak bark ground in a Corona grain mill and buried in a marshy spot near an old pond. I had collected some perennial weed rhizomes and burned them in a Dutch oven placed in the woodstove, and we took turns grinding it. Valerian flowers were juiced in a cast-iron wheatgrass juicer to be fermented for a few weeks. We dried dandelion flowers and froze the mesentery to make a preparation in autumn.

Spring has been less hectic since our CSA has retired. This is our first season without interns or apprentices, those helpful happy hippies who have been dropping in and out of our lives over the past thirty years. We still welcome visitors and usually find plenty of help when we need it. Sharing the farm and trying to teach about it has benefited my own understanding of what I'm doing. I learn from interns and visitors, and feel deep gratitude for the hundreds, probably thousands, who've helped here and supported the farm in countless ways over the years.

The summer vegetables were planted later than usual this year, on a day like the old-timers would choose, with the moon descending and in front of the constellation of Cancer. Three days after planting I harrowed over the seeded rows, and a four-week drought followed. Everything sprouted, except some beets I experimentally planted way too late. I was able to cultivate all the crops, and they look good. There is still no rain in the forecast and we don't irrigate, but that's okay. Our humus-rich soil holds moisture. The drought did help us get up sixty rolls and one hundred bales of spring hay in good condition, but it was thinner than usual. Our hominy corn, Hickory King, was over a foot tall when the cows got out and demolished it, but it's sprouting back. It must like us like we like it.

I made a few hundred tons of compost earlier this spring. Sometimes neighborhood farmers that don't use their cattle manure want me to take it away, which I'm glad to do. Steiner said to regard bringing in fertility as a remedy for a sick farm. Many farms in Appalachia lost topsoil years ago through poor practices. I've never been a believer in using agricultural chemicals, but I believe in grass and clover, and cattle and compost.

Folks show up regularly in early June to help the farm get ready for the solstice festival. We cut dead trees for a big bonfire, work on water systems, and clean the outhouses. The community aspect of the farm comes alive as we approach equinoxes and solstices, our annual celebration times. This land has touched many people deeply. The biodynamic farming method seems to help create a comfortable atmosphere.

Although the farm is economically viable, it's hard to call it profitable because we keep re-investing in it. There are barns to build, fences to fix, roads to grade, machines to repair, and money from farming doesn't go far. Sometimes we make money, other times not. I doubt I've ever made minimum wage by farming. I think a maximum wage makes more sense. Putting a cap on what the top earners are paid would free up a lot of unused capital. As my wealth comes from selling vegetables, I think stockpiled money should also decrease in value, like the shelf life of produce, to encourage people to keep it in circulation. Obviously, I am not an economist. Fitting a well-managed farm into the modern economy, though, is no small accomplishment. Orchards, berries, flowers, and woodland trails may not be economical, but they keep us happy and well fed.

The cattle, necessary as they are, also offer no guaranteed income because the price of beef fluctuates. Steiner's vision of a diversified farm has a lot of moving parts. The cattle rotations suffer when the gardens require attention. But not having livestock is out of the question. As much trouble as they can be, taking care of them connects us with the whole farm. We share the same atmospheric influences and earth elements, and their babies are quite cute.

Our 100 acres of pastures received 120 tons of lime. Balancing the acidic carbonaceous materials with alkaline rock dusts or wood ashes is extremely important. A lot of what I do as a farmer is reflected in various quotes I've encountered. "The use of ash is viewed so favorably by farmers, that they actually prefer it to the manure furnished by their cattle."

I planted buckwheat where the beets failed to sprout. It's a favorite summer cover crop. The spring gardens were thick with wheat, rye, crimson clover, purple hairy vetch, and Austrian peas. It took a bit of work and a lot of patience to mow and incorporate it into the field, but now it is decomposing and adding valuable organic matter. "A field is not sown entirely for the crop, which is to be obtained the same year, but partly for the effect to be produced in the following; because there are many plants which, when cut down and left on the land, improve the soil." Usually, daikon radish and turnips are in the cover crop, and they were, but an unseasonably cold snap last December put an end to them.

Lime helps the clovers to grow. Legumes are essential on our farm, and the ones in our pastures are red and white clover, hop clover, and sericea lespedeza. "Some of the leguminous plants manure the soil, and make it fruitful, while other crops exhaust it and make it barren. Lupines, beans, peas, lentils, and vetches are reported to manure the land." Cows love clovers and we love beans.

Nothing grows well though unless the soil has air, so we keep our gardens well aerated. I have keyline plowed the hillside pastures along the contours to help conserve moisture and add air. In the 5 acres of gardens, chisel plowing and subsoiling deeply last fall allowed the winter rains to enter the soil, and cultivating the crops will continue to keep the soil light and fluffy. Thirty tons of the most beautiful, black, humified compost was spread on each acre. It was made from fermented cow manure, old hay, leaves, and the humus left over from very old, rotten wood chips, along with the preparations.

"Wherein does a good system of good agriculture consist? In the first place, in thorough plowing; in second place, in thorough plowing; and, in the third place, in manuring." Soil needs to breathe. Deep-rooted crops, mulching, resting in pasture or meadow, and even liming can also help keep the soil open. Rototillers beat the soil to smithereens so are not a good option. To bring air into the ground, tools with shanks are better, such as chisel plows, harrows, and small cultivating tractors. "A soil to be fertile must, above all things, be light and friable, and this condition we seek to bring about by the operation of plowing."

These quotes are not from Rudolf Steiner. They are from Pliny the Elder, Varro, Columella, Cato, and Virgil, respectively, all writing over two thousand years ago. People have long understood the importance of crop rotations, manure, ashes, thorough tillage, not plowing land, growing legumes, and resting the fields in cover crops. Good farming principles have been around a long, long time. Steiner grew up with them and reminded us of their importance at a moment when agriculture was at a turning point. His lectures are often regarded as the origin of the organic food and farming movement. Farmers, families, and friends tend land, grow food, and take care of each other, they have for thousands of years, and their tried-and-true methods remain the foundation and essence of good farming today.

Still, by the rotation of crops
You lighten your labor,
Only hesitate not to enrich
The dried-up soil with dung
And scatter filthy ashes
On fields that are exhausted.
So, too, are the fields rested
By a rotation of crops
And unplowed land in the meanwhile
Promises to repay you.

—*The Georgics* of Virgil
30 B.C.

Old-Time Agriculture
Literature from Jeff's Library

The following is a selection of old books and other literature from my library that helped me understand the way people farmed and thought about farming a hundred years ago:

Allen, R. L. *New American Farm Book*. New York: Orange Judd Company, 1895.

Allerton, Samuel W. *Practical Farming*. Rand, McNally & Company, 1907.

Bailey, L. H. *The Nursery-Book*. New York: The Rural Publishing Company, 1891.

Bailey, L. H. *The Principles of Agriculture*. New York: The Macmillan Company, 1906.

Bailey, L. H. *The Pruning Manual*. New York: The Macmillan Company, 1916.

Bailey, L. H. *The Standard Cyclopedia of Horticulture*. New York: The Macmillan Company, 1925.

Bailey, L. H., ed. *Soils: Their Properties and Management*. New York: The Macmillan Company, 1916.

Bailey, L. H., and Walter M. Coleman. *First Course in Biology*. New York: The Macmillan Company, 1914.

Bergen, Joseph Y., and Otis W. Caldwell. *Practical Botany*. Boston: Ginn and Company, 1911.

Biggle, Jacob. *The Biggle Garden Book*. China: Skyhorse Publishing, 2014 (reprint).

Boss, Andrew. *Farm Management*. Chicago: Lyons & Carnahan, 1923.

Brown, Herbert W. *A Living from Eggs and Poultry*. New York: Orange Judd Company, 1917.

Buel, J. *The Farmers' Instructor*. New York: Harper & Brothers, 1840.

Burkett, Charles William, Frank Lincoln Stevens, and Daniel Harvey Hill. *Agriculture for Beginners*. Boston: Ginn & Company, 1914.

Card, Fred W. *Bush-Fruits*. New York: The Macmillan Company, 1909.

Carver, Thomas Nixon. *Principles of Rural Economics*. Boston: Ginn and Company, 1911.

Comstock, Anna Botsford. *Handbook of Nature Study*. Ithaca: Comstock Publishing Company, 1911.

Davis, K. C. *Productive Farming*. Philadelphia: J. B. Lippincott Company, 1922.

Davis, K. C., ed. *Soil Physics and Management*. Philadelphia: J. B. Lippincott Company, 1917.

Downing, A. J. *The Fruits and Fruit Trees of America*. New York: John Wiley, 1855.

Doyle, Martin. *Farm and Garden Produce*. London: G. Routledge & Co., 1857.

Duggar, John Frederick. *Agriculture for Southern Schools (Tennessee Edition)*. New York: The Macmillan Company, 1913.

Fitz, James. *Sweet Potato Culture*. New York: Orange Judd Company, 1911.

Fleet, S., ed. *The New-York Farmer and Horticultural Repository*. New York: New York Horticultural Society, 1829.

Fletcher, S. W. *Soils*. New York: Doubleday, Page & Company, 1908.

Fream, W. *Elements of Agriculture*. London: John Murray, 1892.

Goff, E. S. *Principles of Plant Culture*. Madison: University Co-Operative Co., 1909.

Gray, Asa. *Conversations on Gardening*. London: John W. Parker, 1838.

Henderson, Peter. *Gardening and Farm Topics*. New York: Peter Henderson & Co, 1884.

Henderson, Peter. *Gardening for Pleasure*. New York: Orange Judd Company, 1892.

Henderson, Peter. *Gardening for Profit*. New York: Orange Judd & Company, 1867.

Hensel, Julius. *Bread from Stones*. Translated from German. Kansas City: Acres USA, 1991.

Holmes, Francis Simmons. *The Southern Farmer and Market Gardener*. Charleston: Burges & James, 1842.

Hopkins, Cyril. *Soil Fertility and Permanent Agriculture*. Boston: Ginn & Company, 1910.

International Correspondence Schools. *The Farmer's Handbook*. Scranton: International Textbook Company, 1912.

Jackson, C. R. *Agriculture Through the Laboratory and the School Garden*. New York: Orange Judd Company, 1906.

James, C. C. *Agriculture*. Toronto: George N. Morang & Company, 1900.

Johnson & Stokes Seed Company. *Farm Gardening with Hints on Cheap Manuring*. Philadelphia: Johnson & Stokes, 1898.

Lloyd, John W. *Productive Vegetable Growing*. Philadelphia: J. B. Lippincott, 1914.

Long, Harold C. *Common Weeds of the Farm & Garden*. London: Smith, Elder, & Co., 1910.

Lorain, John. *Nature and Reason Harmonized in the Practice of Husbandry*. Philadelphia: H. C. Carey & I. Lea, 1825.

MacGerald, Willis, ed. *Practical Farming & Gardening*. Chicago: Rand, McNally & Company, 1902.

Martin, John N. *Botany with Agricultural Applications*. New York: John Wiley & Sons, 1919, 1920.

Miller, Claude H., ed. *Garden and Farm Almanac for 1911*. Garden City: Doubleday, Page & Company, 1911.

Roe, E. P. *Success with Small Fruits*. New York: P. F. Collier & Son, 1881.

Sears, Fred Coleman. *Productive Small Fruit Culture*. Philadelphia: J. B. Lippincott, 1925.

St. John de Crevecoer, J. Hector. *Letters from an American Farmer*. London: J. M. Dent & Sons Ltd., 1782.

Tawell, G. H. *Good Market Gardening or the Art of Commercial Horticulture*. London: The English Universities Press Ltd., 1948.

United States Department of Agriculture. *Yearbook of the Department of Agriculture (1859, 1879, 1888, 1893, 1897, 1909, 1912, 1916, 1917, 1924, 1925)*.

Washington: Government Printing Office, 1860,
1880, 1889, 1894, 1898, 1910, 1913, 1917, 1918, 1925,
1926.*

Voorhees, Edward B. *First Principles of Agriculture.*
New York: Silver, Burdett & Company, 1895.

Waters, Henry Jackson. *Elementary Agriculture.*
Boston: Ginn & Company, 1923.

Weathers, J. *French Market Gardening.* London: John
Murray, 1909.

Weir, Wilbert Walter. *Productive Soils.* Philadelphia:
J. B. Lippincott Company, 1920.

Welborn, W. C. *Elements of Agriculture: Southern and
Western.* New York: The MacMillan Company,
1908.

Wilkinson, John W. *Practical Agriculture.* New York:
American Book Company, 1909.

Wilson, John, and W. T. Thompson. "Agriculture." In
Encyclopaedia Britannica, New Werner ed. Akron:
The Werner Company, 1903.

---

* I also have a collection of more modern *Yearbooks of the
Department of Agriculture,* and there is no clearer picture
of the decline of good farming practices and the rise of
chemical agribusiness.

Index

About the Author

Alan Messer

JEFF POPPEN, a Midwestern farm boy, helped develop an organic farm and Tennessee homestead in the mid-1970s, and ten years later began applying biodynamic methods and making the preparations to do so. His livelihood comes primarily from vegetables and cattle grown on the 270-acre Long Hungry Creek Farm, where cows, compost, and community keep the land vibrant and productive. Jeff advocates for a more peaceful agriculture by mentoring young farmers and gardeners, along with a bit of lecturing, consulting, hosting events, and facilitating a few new farm enterprises. His style of old-time farming comes from paying close attention to what elder farmers thought, felt, and did, and by studying how farms were managed before agricultural chemicals were first manufactured on a large scale over one hundred years ago. Like his animals, he gets his food from the farm.